Why Are There Still Creationists?

Why Are There Still Creationists?

Human Evolution and
the Ancestors

Jonathan Marks

polity

First published in 2021 by Polity Press

Polity Press
65 Bridge Street
Cambridge CB2 1UR, UK

Polity Press
101 Station Landing
Suite 300
Medford, MA 02155, USA

ISBN-13: 978-1-5095-4746-3
ISBN-13: 978-1-5095-4747-0(pb)

A catalogue record for this book is available from the British Library.

Typeset in 11 on 14 pt Sabon by Servis Filmsetting Ltd, Stockport, Cheshire
Printed and bound in Great Britain by Short Run Press

The publisher has used its best endeavours to ensure that the URLs for external websites referred to in this book are correct and active at the time of going to press. However, the publisher has no responsibility for the websites and can make no guarantee that a site will remain live or that the content is or will remain appropriate.

Every effort has been made to trace all copyright holders, but if any have been overlooked the publisher will be pleased to include any necessary credits in any subsequent reprint or edition.

For further information on Polity, visit our website: politybooks.com

Contents

Contents

Acknowledgments

This book was way too much fun to write. It was completed while I was a Director's Fellow at the Notre Dame Institute for Advanced Study. I am very grateful to Brad Gregory, Meghan Sullivan, Don Stelluto, Carolyn Sherman, Kristian Olsen, and the rest of the NDIAS, for their support and collegiality. Parts of this book were presented in seminar there, and for participating in the stimulating and helpful discussion that followed, I thank Ani Aprahamian, Dylan Belton, Eileen Hunt Botting, Eric Bugyis, Fr. Terrence Ehrman, David Bentley Hart, Faisal Husain, Robert Latiff, Yulia Minets, Cara Ocobock, Matt Ravosa, Phillip Sloan, and Joshua Stuchlik.

For their especially valuable comments on the manuscript I thank Thomas Bolin, Neil Arner, and Sarah Morice Brubaker.

At UNC Charlotte I have benefitted from the wisdom of my co-instructor Joanne Maguire and my colleague Gregg Starrett. The lunchtime discussions with Mark

Acknowledgments

Pizzato, Trevor Pearce, Bill Chu, Ron Lunsford, and Mike Corwin were also both fun and useful in shaping this book.

It has been a pleasure to interact with biblical scholars and theologians over the past few years, and I am particularly intellectually indebted to Celia Deane-Drummond and Agustín Fuentes and the other participants in their stimulating conference, "Humility, Wisdom, and Grace in Deep Time" back in 2017, which resulted in a wonderful volume called *Theology and Evolutionary Anthropology*. Thanks to my editor at Polity, Jonathan Skerrett, for seeing the manuscript through from beginning to end. Thanks to Karen Strier for decades of insights. For their encouraging notes and comments I thank the reviewers, especially Reviewer #1.

And as always, I am grateful for the support of my wife, Peta Katz, through the creative process and beyond.

Preface

There is a joke that goes, "What's the difference between a biblical literalist and a kleptomaniac?" – "The biblical literalist takes things literally, and the kleptomaniac takes things, literally."

The biblical literalist, however, also rejects what science says about where we came from, whereas the kleptomaniac, or at least the educated kleptomaniac, acknowledges that our bodies and genes are very similar to those of apes, and that a couple of million years ago in Africa, there were no people, but there were apes that had some key human features. The key features were small canine teeth, long thumbs, and a lower body that provided a range of movements like a human's; that is to say, standing up, walking, and running.

A creationist is someone who accepts a literalist reading of the beginning of the Bible in lieu of the scientific narrative that our species has descended from other, earlier species over the course of hundreds of millions of years.[1] There are of course many scholars who understand evolution, and science more generally, to refer to a set of *secondary* causes and processes, while simultaneously maintaining faith in a transcendent *primary* cause, who is in essence God-the-Evolver.[2] Or, as theologian Sarah Coakley puts is, "God is that-without-which-there-would-be-no-evolution-at-all."[3] Whether life is ultimately meaningful is an interesting question, but not a scientific one – since science concerns itself with empirically based inferences, not with spiritual or moral propositions. At issue here is simply whether the origin of people involves apes as ancestors a few million years ago, as the comparative anatomical, genetic, and fossil evidence strongly seems to indicate.

Every generation of evolutionists, however, also inscribes their values into their science. That is not an adulteration of the science, but simply a consequence of being a cogitating social animal. Sometimes those values are sexist (see Charles Darwin's *Descent of Man*, 1871), racist (see Ernst Haeckel's *History of Creation*, 1876), cooperative (see Peter Kropotkin's *Mutual Aid*, 1902), xenophobic (see Charles Davenport's *Heredity in Relation to Eugenics*, 1911), colonialist (see William J. Sollas's *Ancient Hunters*, 1911), egalitarian (see Theodosius Dobzhansky's *Mankind Evolving*, 1961), hereditarian (see E. O. Wilson's *Sociobiology: The New Synthesis*, 1975), or reductive (see Richard Dawkins's *The Selfish Gene*, 1976).

Some scientists try to link their evolution to their atheism. That troubles me, because it makes a positive assertion – "God does not exist" – in the absence of appropriate scientific evidence and inference. Although that assertion is a reasonable hypothesis, I don't think it is mandated by science.

So let me position myself. I am agnostic about God. I capitalize Him out of politeness and custom. But I do not know whether supernatural beings of any sort exist. If they do, that would be nice; and if they don't, that also works. I find it difficult to believe that if they *do* exist, they would care whether or not I *believe* that they exist, when it would actually be very easy to convince me, if they really did exist and care. The only beings that I am aware of interacting with are the ones inhabiting the natural realm, not the supernatural.

I sometimes invoke God, but generally situationally and transiently; for example, towards the waning moments of a Carolina Panthers football game. Usually it doesn't help.

I have no quarrel with people who believe in God, or are generally religious, as long as they don't (1) maintain that their position is validated by science; or (2) try and wheedle me into adopting their beliefs. That directly parallels how I feel about atheists.

I don't think it is "human nature" to believe in God, but I do think it is human nature to think symbolically and imaginatively, rather than resolutely materially.

With that out of the way, let me briefly answer the question posed in the title of this book. There aren't "still" creationists at all. There have always been people who are uncomfortable with the idea that our species is the product of a naturalistic descent from ape ancestors. Christian fundamentalism, which dates to the early twentieth century, mandated a biblical literalist theology, but modern-day opposition to human evolution is actually the product of a reactionary descent from 1960s pseudoscience. In particular, it descends from *The Genesis Flood*, a book first published in 1961, and devoted to the proposition that everything you know about geology and earth history is wrong. Instead, there really was a worldwide flood a few thousand years ago in which Noah and his family and pairs of all the animals were the only survivors. And incidentally, evolution is wrong, because God had created all species *ex nihilo* not long before that.

The intellectual and cultural context of that book is worth considering. As we will note in Chapter 3, just a decade earlier the scientific community had been scandalized by a book that denied and rewrote not biology, but astronomy. It was published in 1950 and called *Worlds in Collision*. Its author was a Russian-born psychoanalyst named Immanuel Velikovsky.

Velikovsky took a classic question from outdated biblical criticism: Falsely assuming that stories are just poorly remembered histories, then what natural phenomena might have been mis-remembered in the Bible as miracles? He then combined his pseudo-biblical musings with his readings of other mythological corpora to arrive at a stunning conclusion: The Hebrew Exodus from Egypt under Moses was accompanied by the planet Venus shooting out of the Great Red Spot of Jupiter, veering close to Earth and causing the biblical Ten Plagues, then careening into Mars, before both planets eventually settled into their now-familiar orbits. Of course, the science of astronomy would have to be refitted to accommodate this bizarre theory.

Needless to say, the scientific community didn't take that at all well, although the astronomers did a famously bad job of trying to engage with and refute *Worlds in Collision*. Their arguments were properly dismissive, necessarily technical, sometimes ad hominem, and occasionally incoherent.[4] And although Velikovsky's ideas eventually receded from public consciousness, there were significant parallels between *Worlds in Collision* and *The Genesis Flood* scarcely a decade later. Both prominently cast themselves against science, and in favor of their particular interpretations of the Bible. One bluntly opposed astronomy, the other geology. Yet the biblical text figures prominently in both, as misunderstood "history" in the colliding planets narrative, and as properly understood "history" in creationist narratives.

We have engaged most commonly with biblical literalist creationism as a false theory of biology,[5] or as an archaic remnant of older modes of thought;[6] but it is modern, not primitive,[7] and treating it as a false story simply replicates the astronomers' frustrating engagement with *Worlds in Collision*. It will always prove unsuccessful to engage with creationism in terms of "our story is true and yours is false" – since, at the very least, many aspects of *any* story of human evolution are debatable or downright inaccurate. Indeed, both evolutionist and creationist narratives of human origins have at times freely incorporated racist elements.

The thesis of this book is that modern creationism is not part of a vast conspiracy of stupid. It indeed opposes the normative views of science, but that opposition is different from the economic roots of climate-change denial, the misguided yet still unbiblical sincerity of the anti-vaccinators, or the sheer perversity of the flat-earthers. Of these popular modern "anti-science" positions, only creationism is religiously motivated. It is consequently a special kind of anti-science. To grapple effectively with creationism, then, the scholar of human origins and the scholar of religion are natural allies.

Happily, those two scholarly endeavors converge in anthropology.

This book will adopt two positions about religion and science, or more specifically about evolution and creationism, which seem unfortunately uncommon but are nevertheless rather straightforward and true. First, one can take the Bible seriously (as sacred writings, as literature, as a glimpse of ancient life, as ancient wisdom) without taking it literally. Second, most Catholics, Jews, and even Protestants aren't literalists. Consequently, to the extent that this is a scientific and a religious issue, it isn't science vs. religion. It is religion vs. religion about science. By implication, then, the argument between evolution and creationism is ultimately a sectarian theological dispute within Protestantism (even Islamic creationism is derived from the Protestant literature), and consequently the appropriate battleground is not science at all, but theology. Science, especially biology, is marginal to the question of whether the Bible should be taken literally.

I
Introducing the Ancestors

It is not a secret that about half of Americans are morons. Were the journalist H. L. Mencken alive today, he would very likely regard that as a considerable understatement. They eschew vaccinations. They take right-wing provocateurs seriously. They vote against their rational interests. They can't distinguish between gut feelings and informed thoughts, and privilege the former over the latter when they can. And they aren't all necessarily even the same people.

There is a veritable industry of aggrieved social critics condemning the stupidity of ostensibly modern citizens who reject science. But of course nobody totally rejects science, and maybe they have some reasons for rejecting some particular science. After all, not all science is good. Back in the 1920s, when the science of the age called for solving social problems by sterilizing the poor and restricting the immigration of genetically feebleminded Italian and Jewish immigrants, the people who were anti-science were actually in the right.

Why Are There Still Creationists?

We all make decisions about what science to accept, what science to ignore, and what science to reject. You probably don't give much thought to helminthology, the science of parasitic worms, generally found in feces. Perhaps, like me, you don't give much thought or credence to exobiology, the science of non-existent extraterrestrial life. You may never even have heard of quantum electrodynamics, but it sure sounds scientific.

This book is about the rejection of evolution, a science more real than exobiology, more familiar than quantum electrodynamics, and more decorous than helminthology. Evolution is the science of where we come from, a question so basic to human existence that all peoples have stories to answer it. It is about ancestry, and the framework of this book rests upon an anthropological truism: The ancestors are always sacred.

Confronting this cultural fact will help explain not only the popular rejection of human evolution, but the often bizarre and vituperative disputes within the science itself. In the 1980s, for example, scholars working on *Homo habilis* in Kenya fought bitterly with scholars working on *Australopithecus afarensis* in Ethiopia over whose fossils were more important to the reconstruction of human prehistory. Today the Kenyan and Ethiopian fossils are reconciled and have joined forces against the upstart fossils from South Africa (*Australopithecus sediba* and *Homo naledi*). Regardless of the zoological reality of these species, they are the subjects of mythology and nationalism, not to mention fame – which is why zoological realities and paleoanthropological realities don't necessarily map on to one another well. The fossils are national treasures, and the species they represent are, in the broadest sense, sacred ancestors.

Consequently, when scholars reject *"Australopithecus prometheus"* or *"Homo rudolfensis"* as unreal, what they mean is that the fossils allocated to them ought to be called something else. Making scientific sense of the ancestors is no small undertaking.

Neither is making unscientific sense of the ancestors. When evangelical entrepreneurs build Noah's ark in Kentucky, they must still struggle to reconcile biogeography and adaptation, as did serious scholars two centuries before them. If God made animals to fit where they live – polar bears to the arctic, bison to the Great Plains, koalas to eucalyptus forests – then it seems unlikely that they could have gotten there from Mount Ararat before going extinct, without some other *ad hoc* miraculous intervention. The problem actually arises from juxtaposing two different ancient sources. The Bible talks about Noah's ark, but not about adaptation; that is from Aristotle. Moreover, if God made birds to fly, then it is hard to explain ostriches and penguins. Those are the kinds of facts that led scholars 200 years ago to begin to seek explanations in history, rather than in miracle.

It doesn't matter whether you call the animal-saver Noah, or Deucalion, or Utnapishtim – as the stories from various ancient sources had it – he still had an impossible job of dropping off the various animals in their respective habitats around the world. He must be tasked with dropping the lemurs off in Madagascar, anacondas in the Amazon, armadillos in America, gibbons in Southeast Asia, kangaroos in Australia, reindeer in northern Europe – or else another story needs to be composed and related about how they got there. And that story is necessarily non-biblical – because the Bible doesn't say anything about it, which in turn undermines the assumption that the Bible is giving us a complete and accurate account of adaptation and biogeography. In fact it undermines the assumption that what the Bible says is even relevant to understanding the facts of pre-history. Early scientists 200 years ago recognized the problems here.

Fossils made the problems with Noah's ark as a scientific explanation for the patterns of life even more acute. Some ancient animals seemed to cross-cut categories – large swimming and flying reptiles, for example – while others seemed to confuse familiar patterns. Elephants, after all, were only known to be from hot climates, so why were there woolly elephants in Siberia long ago? Certainly being woolly helps you survive in Siberia, but these were elephants, so what was their relationship to the normal elephants of Africa and South Asia? And what happened to them? Was God's gift of woolliness somehow insufficient for them? And why didn't the same thing happen to the non-woolly elephants of the tropics? The science of the early 1800s was discovering that the world of the past was a different place than the world of the present, and the Bible afforded no guide to understanding it.

Were there pteranodons and iguanodons in Eden? If not, then how did they get into the fossil record? And if so, then why doesn't the Bible mention them? The Bible implies quite directly that Eden was populated by familiar creatures. As a character in *The Sopranos* articulated it many years later, "No way! *T. rex* in the Garden of Eden? Adam and Eve would be running all the time, scared shitless. But the Bible says it was paradise." And clearly, if you imagined them as somehow uninterested in consuming Adam and Eve, those T. rexes would have had to be the world's worst adapted herbivores, about as suited to vegetarianism as great white sharks.

It is no coincidence that biblical scholarship and biological scholarship matured together. The coevolution of information and explanation had been a long-term process. Scholars in ancient times had envisioned the relationship between God and His creation as analogous to that of a king and his subjects. He ruled by decree, and could reverse or abrogate his decisions more or less capriciously. For example, God devotes a chunk of space in Leviticus and Deuteronomy to explaining what foods are clean and unclean for His followers. But in Mark 7:19, Jesus peremptorily declares "all foods clean." In practical terms, that certainly made it easier to be a Christian than a Jew, but it also reveals a God who seems not to be able to make up His mind.

Could the sea part for His followers and swallow up their enemies? Maybe. Could a man live for three days in the belly of a fish? Maybe. Could the sun stand still in the sky for twelve hours? Maybe. Can the dead be raised? Maybe.

By the eighteenth century, however, the image of God as a sort of cosmic despot was being gradually supplanted by the image of God as a sort of cosmic engineer, building stable things that work a certain way, always. The universe now was seen to run according to natural laws, which were inviolable by their very virtue of being divine, which in turn gave less leeway to suspend their operation, for that would plunge the universe into chaos, which was precisely what the Creation had transcended.

To a large extent, this was a consequence of the

discoveries of physics and astronomy, which had shown convincingly that the earth moves, that it is a planet like others, that it revolves around the sun, and that it does so because of gravity, which is also what keeps the moon revolving around the earth and reciprocally causes the tides upon the earth. And just as laws keep the moon circling the earth and earth circling the sun, so too do laws keep blood circling through the body. Laws were there in physiology as well as in physics.

Laws are order. Violating them introduces disorder. The earth could not stop rotating for a day without sustaining catastrophic consequences resulting from inertia; and starting it up again would engender similarly daunting implications now predictable from the laws of physics.

Rationalism, the emerging ideology of the eighteenth century, was a powerful antagonist against miracles. It deployed an old medieval weapon in order to define miracles out of existence: If we can explain things without miracles, then why bother with them at all? As for the recollections of the ancients, which explanation is more likely – that there was a temporary suspension of the laws of nature, or that somebody, somewhere along the line, didn't get the facts exactly straight in relating the story?

On this basis, an Enlightenment savant like Thomas Jefferson could start distinguishing the biblical things that Jesus probably did say and do from the biblical things that Jesus probably didn't say and do. And by the 1840s, biblical scholars were using history and linguistics to reimagine the gospel story as non-miraculous – controversially at first, but ultimately with irresistible intellectual force. After all, French and Spanish could

not have arisen miraculously at the foot of the Tower of
Babel if they had been spun off from Vulgar Latin over
the past 1,000 years or so.

But of course there was much more going on in
European and American intellectual life than the de-
miraculizing of Scripture. Miracles were being written
out of earth history as well, being supplanted by "uniformitarianism," which held that modern geological
processes are generally slow and gradual, and are the
processes that are most appropriate to apply to try and
understand the earth's past. And when you performed
that application, the earth seemed to be far older than
the biblical "begats" could allow. The closer you looked
at the composition and patterns of geological formations, the more it seemed as though the earth seemed
to have "no vestige of a beginning."[1] The remains of
ancient life embedded within the geological formations
indicated primeval worlds inhabited by only remotely
familiar forms of life. Indeed, the history of life was intimately bound up in the history of the earth itself.

Rethinking Species and Scriptures

By the 1840s, then, two large questions loomed over
the closely allied fields of natural history (documenting
the diversity and history of life) and natural theology
(trying to make sense of the diversity and history of
life). In natural history, two facts had been well established. First, species have gone extinct. That is to say,
the termination of species was acknowledged, although
without any biblical justification. Extinction was, after
all, what Noah was specifically charged with prevent-
ing, and the biblical narrative that we have certainly

indicates that he succeeded. Second, different species lived at different times, correlating well with the geological formations in which they were embedded. Reconsidering the extinct and extant animals he had seen in South America, an English naturalist wrote in 1845, "This wonderful relationship in the same continent between the dead and the living will, I do not doubt, hereafter throw more light on the appearance of organic beings on our earth, and their disappearance from it, than any other class of facts."[2]

From the standpoint of natural history, then, if species lived and species ended, then it was only natural to theorize how they might originate. That was the natural history question.

Simultaneous respect for data and Scripture, however, created a problem for mid-nineteenth-century scientists, facing evidence that showed the earth to be old and species to have lived at different times. The transmutation of species itself was not a particularly new and threatening idea in the 1840s. It had been proposed by Enlightenment scholars in England and France, and was now (in 1844) the subject of a bestseller called *Vestiges of Creation*. But those theories of evolution were theories of progress, in that the change of one species into another was considered to be somehow an improvement. To an age that usually still saw humans atop a line comprising all other earthly species – a Great Chain of Being – this early evolutionary theory essentially involved a short ride on a Great Escalator of Being.

But there was no evident mechanism for such a drive to improve, aside from imagining that all animals deep inside wish they were human. Nor was it clear that all species could be naturally placed in a linear sequence at all – bears, goats, and chipmunks all seemed pretty much equidistant from people.

So the scholar of 1845 had two unsatisfactory theories to explain the origin of species – biblical creation, or evolution via a Great Escalator. But there was a transient third theory, which recognized the age of the earth and the succession of life, yet tried to remain pious by imagining the origin of species at different times in the deep history of life on earth to be miraculous, not naturalistic. This "non-biblical creationism" was in fact a theory of choice for many of the leading biologists of the age. Indeed, although we tend to see Victorian creationism through the lens of modern creationism, it was principally *not* biblical literalist creationism that pushed back initially against Darwinism, but non-biblical creationism. The leading anti-Darwinians – Richard Owen in England and Louis Agassiz in America – were committed to the age of the earth and the succession of life, but they clung to the origin of species as a series of miracles.[3] They certainly didn't believe in a six-day creation a few thousand years ago as a viable alternative to Darwin's proposal.

Yet non-biblical creationism had its own baggage. In the first place, it was pious but un-biblical, so what really was the point of the piety? And second, where biblical creationism invoked one great miracle as the source of all species, this set of theories invoked lots of little miracles throughout earth history. Yet favoring a theory that invokes many miracles over just one is a hard sell, for miracles are supposed to be rare. That's why we call them miracles.

What non-biblical creationism did, though, was to preserve a critical part of the Christian Genesis story

– the Fall of Man and the need for redemption from sin. That was indeed the point of the non-biblical piety. Perhaps the bit about sin and redemption could be retained if we just imagined the Garden of Eden to have existed not near the beginning of time, but near the beginning of history. Traditionally, Eden was considered to have been inhabited by modern kinds of creatures, which were all that the classical writers knew. But now there was clearly a long period of time when there were pre-modern creatures – wherever they came from – on a pre-modern earth, yet all known evidence of humans was still of a more-or-less modern form. The Garden of Eden, then, might have been the stratigraphic frosting that God slathered upon the top of the geological layer cake, so to speak – the last literal bit of His creation. This would preserve the most meaningful parts of the story – about sin, death, and the need for redemption – while taking most of the biblical creation narrative figuratively, and yet would still permit biblical piety and its companion, biblical morality, to coexist and persevere along with the progress of science. That is, unless incontrovertible evidence arose showing that people – or the tools they left behind, since who else makes stone tools? – coexisted along with extinct animals. Not necessarily with T. rexes, but with anything that was big and scary and no longer around, like woolly mammoths, or sabertoothed tigers, or giant European cave bears. For that would indicate that there was no break at all between the ancient and modern worlds, that the two bleed into one another, and that there could be no place for a literal Garden of Eden.

And yet this was precisely the evidence that continued to mount throughout the 1840s: Some kind of ancient people lived alongside some kind of extinct animals a long time ago. The big question for natural theology, then, concerned the occupation of Eden, the last remaining part of the Genesis story, once you have acknowledged the age of the earth and the succession of life.

Both the natural theology and the natural history questions were ultimately resolved in the same year, 1859, with Charles Lyell's address to the British Association for the Advancement of Science in September, which publicly acknowledged the coexistence of stone tools and archaic animals, and with the release of Charles Darwin's *On the Origin of Species* in November, which theorized the beginnings of species. The following year, a group of liberal theologians published a runaway bestseller called *Essays and Reviews*, which brought modern critical biblical scholarship to the reading public, and passingly referenced Darwin.

All of which is to say that the theological and biological revolutions were closely intertwined. Indeed, just four years later three influential books came out: Thomas Huxley's *Man's Place in Nature*, the first book on human evolution; Ernest Renan's immensely popular *Life of Jesus*, which represented Jesus as a smart blond guy, not as the miracle-working Mediterranean demigod that readers had grown up with; and Lyell's *The Antiquity of Man*, which reviewed the evidence for primitive people in an ancient world, full of extinct and unfamiliar species. If we could no longer distinguish the archaic, pre-modern earth clearly from the modern earth, and the modern earth appeared to be an outgrowth, or an "evolution," from the pre-modern earth, then Eden may well have been populated by archaic species and even by archaic people. In fact, it might not even have been a garden at all, but a glacier or tundra or savanna.

The Bible itself could then be understood as a set of sacred stories, part of a larger group of such stories, but requiring contextualization among the myths and legends of the world, and accessible through the study of ancient histories, languages, and lifeways.

In addition to the development of biblical studies, natural history studies, and prehistoric archaeology, yet another strand of uncertainty was entwined in mid-nineteenth century Euro-American intellectual life – namely, the relationships among living groups of people, particularly slaves and slavers. Were they all of one flesh, and presumably therefore of one origin, as the Bible had it – but which in turn implied considerable mutability of appearance in human face and form since the Garden of Eden? Or might they have been the products of separate creations at different times, distinct since their beginnings – which seemed more harmonious with the new geology, and was also distinctly un-biblical? This debate also encoded a moral argument: if humans were the products of separate creations, then owning a slave might be no different in kind from owning a horse. But if they were the products of a single creation, and we are all ultimately brothers and sisters, then some people regarding other people as property may not seem quite right. And if the ancient earth seemed to lend scientific credence to the possibility of archaic, pre-Adamic racial origins, nevertheless the interfertility of human populations seemed to lend scientific credence to the biblical position.

A more curious biological/theological/political question lay just alongside that of the origin of the human races; namely, the nature and attainment of civilization. Assuming it to be readily enough identifiable, why do some peoples have it and not others? Is it due to a constitutional defect? Or might it be possible to lead people who do not have civilization to it?

By the mid-nineteenth century, exotic peoples were known who did not even practice agriculture but subsisted on wild foods alone. Moreover, they lacked the art of metallurgy, and used only tools made of stone. Archaeologists were also identifying an ancient past when the only tools Europeans themselves used were sharpened rocks – a "Stone Age."

And yet, even this simple cultural history was difficult to reconcile with the Bible. Some patriarchs may have lived before Tubal-cain invented metallurgy in Genesis 4:22, but none was a hunter-gatherer; Adam and Eve had been horticulturalists from the very beginning, as Genesis 3:15 specifies that Eden was there to be tilled. Its authors could not even conceptualize a pre-agricultural human existence.

This, in turn, raised the question about the origin of hunter-gatherers. Were they indeed the most primitive form of humanity – savages, not even having discovered plant and animal domestication (a discovery which would make them barbarians, a step up)? Or were they degenerate descendants of a primordial Edenic horticultural society? Both alternatives are non-biblical, since the Bible doesn't say anything at all about hunter-gatherers, but the degeneration theory was the one that seemed a bit more pious in the mid-nineteenth century. It was at least consistent with the Adamic narrative, as with the "fallen state of Man" theological doctrine.

Unfortunately it was less consistent with the secular doctrine of progress, and with the data from archaeology and ethnography indicating that agriculture came late, and foraging early, in human prehistory. Our species just seemed to have been advancing – perhaps not uniformly or evenly, but advancing nevertheless – from early savagery through barbarism and into civilization, all driven by the engine of technology, the increasing mastery of people over nature, and over each other.

Thus, theology coevolved with science in the nineteenth century. There is a commonplace view which holds that a crucial difference between science and religion is that religion is rigid, but science changes. Thus, a "fundamental difference between religion and science is that the former is all about the celebration of certainty, whereas the latter is all about the quantification of doubt."[4] But that simply isn't true. Not only does "religion" change over time, but "science" can be pretty darn dogmatic as well. They are not at all so readily separable that way.

Theology and Evolution

There is in fact a rich literature by modern Christian scholars on the meaning of evolution for modern Christian life, or on being a Christian in a post-Darwinian world. They all agree that denying evolution is a stupid way to tackle the problem. Smart ways generally invoke God's action through evolution, and generally try to unpack the many possible meanings of "create." As one eminent Christian theologian recently put it:

creation science is largely ignored in most mainstream contemporary theology, which is much more interested in what the doctrine of creation says theologically about the world and the place of human beings in it ... The reason it is rejected in theology is not primarily because it is bad science, but because it is bad theology: In particular, it tends to assume a competitive relation between divine action and natural secondary causation, such that God and nature are taken to be alternative possible explanations of events, thereby denying the immediate dependence of all creation on the Creator for the gift of its existence.[5]

If your goal is to make the natural realm meaningful by recourse to a supernatural realm that is inaccessible to science, then you might as well regard them as complementary, rather than as antagonistic.

But if the history of life does have meaning, unfortunately scientists are not the people to ask about it, because they know about the history of life, but not about its meaning, since that is a question not for science, but for semiotics and metaphysics. 'Metaphysics' is a word that is usually articulated scornfully by scientists, most likely, as a philosopher once noted, because they are afraid of having their metaphysics questioned.

There have been, nevertheless, thoughtful approaches taken by Christian scholars toward rendering evolution meaningful. One influential trajectory was followed by the Jesuit paleontologist Pierre Teilhard de Chardin (1881–1955). Teilhard worked with a teleological theory of evolution – seeing the emergence of our species as the unfolding of a cosmic plan – a genre of theory that is generally rejected by science but is nevertheless frustratingly impossible to disprove. Teilhard then went on to describe the details of that plan, involving the co-evolution of life, mind, and spirit (which we generally tend to treat separately in science, except for spirit), and their ultimate optimistic realization at the mystical Omega Point sometime in the distant future. The eminent evolutionary geneticist Theodosius Dobzhansky found Teilhard's spiritual thought intriguing. On the other hand, the eminent paleontologist George Gaylord Simpson (himself the son of missionaries) once told me it was the only substantive matter he ever disagreed with his great friend Dobzhansky about. But some theologians find Teilhard's evolutionary ideas valuable; he was quoted at length, for example, by Bishop Michael Curry in his sermon at the wedding of Prince Harry and Meghan Markle (see endnote 4, Chapter 4).

Another influential approach was taken by the philosopher Alfred North Whitehead, who, for a non-believer (and erstwhile collaborator with the greatest atheist of all, Bertrand Russell), certainly expended a lot of effort on thinking constructively about God. Whitehead invites you to imagine a universe continually being created, and full of possibilities. In this universe the units of nature are not objects, but transformations; not beings, but becomings. God may nudge you toward certain ones, but this is a universe fundamentally in flux, and its Creation marked not the end but rather the beginning of creating. As a modern theologian asks rhetorically, "But what if God is not just an originator of order but also the disturbing well-spring of *novelty*? And, moreover, what if the cosmos is not just an 'order' . . . but a still unfinished *process*? . . . And suppose also that God is less concerned with imposing a plan or design on this process than with providing it opportunities to participate in its own creation?"[6]

There are, of course, many other intellectual paths toward reconciling Christianity and evolution. One obvious route is to parse the word "create" – particularly when you consider that it can mean various things simply in English. One can create a book by composing it or by printing it, which are quite distinct activities. One can create a diamond by crystallizing carbon, or by cutting and polishing a stone. One can create a pot by molding it into shape, or create a riot by inspiring others into action. One can create a monster literally (like Victor Frankenstein's creature) or metaphorically (like Rupert Murdoch's Fox News). And that is without even considering what nuances might exist in the Hebrew or Greek cognates.

And once you settle on what "created" might mean, you can start thinking about what the Bible might mean by the phrase "in the image of God." Indeed, this has been fertile ground for theological discussion for centuries. The modern theologian Wentzel van Huyssteen argues for thinking about the *imago Dei* "as having emerged from nature by natural evolutionary processes."[7] Theologian/biologist Celia Deane-Drummond draws on evolution and the *imago Dei* to develop a multi-species morality for the modern age.

Clearly Christianity and Darwinism are not antithetical, although there are also prominent scientists who willingly take the bait, elaborate on their own weirdly joyless metaphysics, and proclaim that life is meaningless, there is no God or ultimate justice, "just blind, pitiless indifference" in the universe.[8] So there! Obviously this might be true, but to confuse it with a scientific statement is certainly a mistake. Rather, it is an offensive salvo in a frustratingly defensive war over our ancestry. As the sociologist Bruno Latour acknowledges, "it is certainly a great pity that the only religious minds that neo-Darwinians ever encounter come from creationism."[9]

Ancestry and Relatedness

There is a broad intellectual frame available to us to help make sense of the vigor and longevity of the rejection of evolution. That frame is kinship, the sense people make of their place in a social and moral world by conceptualizing their descent and relatedness to others. This is a particularly human thing to do, as the apes do not (as far as we can tell) have relationships homologous to spouse, in-law, father, grandma – much less kissing cousin, baby daddy, heir, adopted child, step-child, or remote ancestor. These relationships are what structure the course of our lives; for a notable example, in *Game of Thrones* Jon Snow is reminded continually that he is Ned Stark's bastard son (before tragically falling in love with his aunt).

In case of human evolution, though, the passion is focused on whether our ancestors were apes. The ancestors, as noted earlier, are always sacred, in the broad anthropological sense of "special." If you think the apes in our ancestry aren't special, try denying them to a biologist. Benjamin Disraeli made the options clear in 1864: "Is man an ape or an angel? My lord, I am on the side of the angels. I repudiate with indignation and abhorrence the contrary view, which is, I believe, foreign to the conscience of humanity."

Introducing the Ancestors

Of course as a prominent Jew-turned-Anglican, Disraeli was particularly sensitive to the presumption that our descent is a primary determinant of who and what we are. (Indeed, a few decades later, Franz Kafka would use the ape-turned-human as a metaphor for the assimilated European Jew in his story "Report to an Academy.") Disraeli wanted us not to be descended from apes, so that we may be angels. He articulated this crude but widespread ultimatum only a few years after the publication of *The Origin of Species*, when the idea of ape ancestry was still fresh. But why this idea should necessarily preclude our siding with the angels is unclear. Certainly Disraeli himself did not become Earl of Beaconsfield on account of his ancestry, but, if anything, in spite of it.

Our similarity to the anthropoids was in fact already well known. An old Roman poet named Quintus Ennius observed how similar we are to the monkey, the most horrid of beasts ("Simia quam similis turpissima bestia nobis"), a saying preserved by Cicero and repeated in the foundational scientific works of Francis Bacon and Carl Linnaeus. Disraeli could hardly deny the similarity of human and ape, but what vexed him was its meaning – a question that had indeed long vexed scholars. What is the meaning of a creature who matches us physically bone for bone, muscle for muscle, organ for organ, and yet is not human? That, in fact, had been the question initially posed by the anatomist Edward Tyson back in 1699. Knowing that walking and talking served to distinguish us from the animals, Tyson was fascinated by a creature recently brought back from Africa. Its position could readily be established by its lack of language, which was intimately connected to thought, morality, and spirit. This creature – which we now know to have been a chimpanzee – lacked the power of speech, and thereby lacked the connection to God that all humans have.

On the other hand, there did not seem to be any obvious anatomical reason (for the state of knowledge in 1699) why the creature should not be able to walk like us. Its body corresponded eerily to our own, and yet Tyson had not actually seen the creature walk. Rather, it used the knuckles of its hands to bear its weight. That must have been, reasoned Tyson, because it was mortally ill (which it was), so Tyson compromised and had the creature drawn standing up, with the aid of a cane. The chimpanzee was thus anatomically continuous with people, but intellectually (and presumably spiritually) a gulf away.

This anatomical proximity between human and ape became increasingly familiar over the course of the eighteenth century, but remained difficult to explain. By the mid-1700s, the naturalist Count de Buffon had recognized that the most obvious meaning of this similarity, indeed of patterns of physical similarity in the animal kingdom generally, was that it signified a trail of descent. But once you saw it like that, where would such a trail end?

After all, Buffon reasoned, if you argued that the donkey is built so much like a horse that it should be considered part of a horse "family," and is thus some sort of transformed primordial horse, then two things followed. First, you could say exactly the same thing about human and ape; and second, you couldn't stop there, because you could make precisely the same argument for equids and hominids being transformed variations of a mammalian "family," or mammals and birds being transformed variations of a vertebrate "family." The only clear stopping point would be the origin of life itself. By the same argument that God made a primordial feline, from which lions, ocelots, and panthers are all descended, one could argue just as readily that God made a primordial mammal, from which lions, ocelots, panthers, bears, giraffes, monkeys, and beavers are all descended; or a primordial vertebrate, from which lions, ocelots, panthers, bears, giraffes, monkeys, beavers, eagles, frogs, and tunas are all descended. This was a scary intellectual place – in which all life might be genealogically related by virtue of common descent, as indicated by their nested patterns of anatomical correspondence – and Buffon retreated from it. A century later, Darwin eventually proved him right.

2

Scientific Stories of Our Ancestors

Linnaeus showed in the eighteenth century that, as living organisms, humans are most similar to apes, and a classification of animals based on their similarities ought to reflect that fact. A century later, Thomas Huxley argued – based on Darwin's recent insights – that our "place in nature" was deducible from our ancestry, and that the apes provided the best evidence of that ancestry.

The apes afford both continuity and contrast to the human condition. Huxley himself recognized the dialectical nature of the relationship between us and the apes: "[N]o one is more strongly convinced than I am of the vastness of the gulf between civilized man and the brutes; or is more certain that whether *from* them or not, he is assuredly not *of* them."[1] Thus, in constructing a scientific narrative of our origin, we begin with the life of apes, and reconstruct how that became differentiated into the life of people.

Our first recognition must be that the most familiar aspects of our lives – of the life of what Huxley called "civilized man" – are the product not so much of biological evolution as of social history. The history of food production, to be precise. From believing in anthropomorphic deities to living in cities, to having jobs, to patriarchy and inequality – these aspects of modern life are all, in some fashion, historical consequences of producing and controlling food, storing it up, doling it out, defending it, and, most importantly, knowing it will always be there, rather than passively relying on the earth, sea, sky, and weather to provide it. To identify the evolutionary, or biological, distinctions between human and ape, we have to unite the post-industrial representatives of our species with the farmers and foragers of our species, and juxtapose them against the apes. While this may be obvious to us now, it was not so obvious to the early Darwinian spokesman, Ernst Haeckel, who explained: "If one must draw a sharp boundary, . . . it has to be drawn between the most highly developed and civilized man on the one hand, and the rudest savages on the other, and the latter have to be classified with the animals."[2] The reaction from his English colleagues was that "at least he's on our side" – "our" side being the side of evolution, not necessarily that of humanity.

Grouping humans together, no matter how rude their state of existence, in a contrast with the apes, allows us to see human evolution differently than Haeckel did. We will see three general features that differentiate us from the apes: our manner of locomotion, our manner of communication, and our manner of interacting with one another. The first of these is the adoption of a bipedal stride, walking or running more slowly than an ape, but in a manner that the ape can only do for short distances, and clumsily. The second is language, with its attendant physical aspects and cognitive functions, generally referred to as "symbolic thought," producing and attributing meaning to things and sounds largely arbitrarily. The third involves the creation and adoption of abstract rules to establish networks of reciprocal obligations that join people together as families; thus it is not so much an evolved property of humans, as an evolved set of relationships among humans.

Human Ancestry as the Dental and the Mental

It does seem that by 4 million years ago there was at least one, quite possibly several, kinds of apes that were moving around on two legs (detectable in the anatomy of their pelves, knees, and feet), quite differently from other apes, and who had small canine teeth, a prerequisite for speech. Having small, non-sexually-dimorphic canine teeth is a condition quite unique among the apes, and suggests different modes of competition among males and mating criteria among females. They still had the brains and arms of apes, however, and weren't transforming significant aspects of their environment into tools in any obvious way.

We group these creatures into the genus *Australopithecus*, and, like Haeckel, we would probably see them as apes, with key aspects of locomotion and sound production that place them on our particular branch of the family tree, yet leading essentially ape lives. This would be where we place the famous fossil from the 1970s, Lucy, who lived over 3 million years ago. The genus level of our classification is most useful here, for the taxonomy of our own lineage is not quite like the taxonomy of other zoological lineages: there are unusual issues of individual recognition and national pride in the scientific mix. The practice of naming the ancestors is invariably sacred, even in science; consequently, it is not really worth asking whether *Australopithecus afarensis, africanus, bahrelghazali, prometheus, garhi, sediba,* and *deyiremeda* are "real," or whether their subtle differences actually ought to be comprehended within just, say, two or three species. Fossils do not come with species labels; they are assigned species labels. The fossils are "real" as material objects to be identified and named, and the species labels are "real" as linguistic markers. Their "reality" lies not in their status as biological species, however, but in their status as narrative elements in a story about our origins, that is, as mythic ancestors. This is not to say that there were no species or real ancestors; simply that they inhabit a different reality than that of the species *Drosophila melanogaster* (the fruit fly) or *Ursus spelaeus* (the extinct giant Eurasian cave bear), and it is a mistake to try and understand the species names in our own lineage as if they referred to the same kinds of things as other biological species names do. That is why biologists have traditionally had such a difficult time grappling with the taxonomy of our lineage.

In addition to not marking themselves with species identifiers, our ancestors also did not mark their descendants for us. Ancestor-descendant relationships are always inferred, not discovered; what the fossils yield are patterns of relative similarity and relative chronology. We observe two fossils that look alike and lived close in time, and we infer that something very much like the earlier form gave rise to the later form. In this manner, we see that the descendants of *Australopithecus* had bifurcated by about 2 million years ago. One line of descendants exaggerated the dental and masticatory specializations of *Australopithecus* – de-emphasizing the front teeth and emphasizing the back teeth – and became the genus *Paranthropus*, retaining the bipedal adaptations of *Australopithecus*, keeping the brain small and surviving with its massive molars. The other genus also retained the bipedal adaptation of *Australopithecus*, but eschewing the chewing adaptation for cranial adaptation, becoming the genus *Homo*, and surviving with its massive brain.

Just as we do not know what bipedalism was "for" – it doesn't seem to be an obviously superior way to get around, although many possibilities have been suggested – we likewise do not know what our big brain is "for." We can observe what it does: it permits us to explore and exploit our environments in creative ways; to solve problems in the real world; to envision things that don't exist, or might exist, or ought to exist, in an imaginary world; to live our lives in accordance with abstract and arbitrary rules. One other important thing our brain does is force us to communicate in a zoologically unique way, symbolically.

Some animals communicate principally with clicks and whistles, others by scents, and still others by movement. We do it by vocalizations, deployed in a unique way, based on largely arbitrary sets of learned rules, which give local meaning to particular sounds (phonemes), to combinations of sounds (vocabulary), to organizations of the sound combinations (syntax), and to the context, manner, and tone in which the organized sound combinations are delivered (pragmatics). It is analytically useful to distinguish the mechanisms of sound production (speech) from the cognitive processes behind it (language), although obviously the vocal apparatus and the cognitive apparatus coevolved, because they presently work together – and somewhat differently than they seem to work in apes. Nevertheless, we know very little about the processes by which the human mind came to assign locally specific meaning to sounds, words, and utterances. Despite many decades of attempts to teach apes to communicate with us in various ways, it seems as though they do not actually have anything of interest to say to us, and are no better at language than they are at bipedalism. That is to say, it simply isn't what they evolved to do.

The Revolution in Communicating, Interacting, and Thinking

Language was a sufficiently powerful new means of communicating that it seems to have evolved in spite of the liabilities it posed for our ancestors. Perhaps having small canine teeth permitted language to evolve in our lineage; an ape's ability to make distinct, intelligible sounds is limited by its large, interlocking canine teeth. But in being able to make such sounds, without those large canine teeth, our ancestors also lost the classic primate defensive weaponry. Moreover, our throat's anatomy is slightly different from an ape's, with the net effect of producing new sounds, but also making it far easier for us to choke. Our tongues are also co-opted so that they can no longer do efficiently what they do in apes – namely, dissipate heat by panting. Our ancestors' bodies thus had to find another way of dissipating heat, and came up with evaporative cooling: our skin has a higher density of sweat glands than an ape's does. But since evaporative cooling only works when the skin is exposed to air, our body hair had to degenerate into thin wisps (yet retaining the same density of hair follicles as an ape's skin).

But the most important compromise our ancestors had to make in evolving language came as a consequence of simply how language works: It is learned, slowly. After being born, we cannot communicate properly for several years. This biological investment in learning shows up in two fairly subtle (as all such features are) but significant differences from the apes. First, we have an extended period of immaturity relative to an ape; for example, a human is about twice as old as a chimpanzee when their wisdom teeth erupt. And second, our newborn brains are considerably larger than an ape's. Unfortunately, the human birth canal hasn't quite kept up the pace. Consequently, an ape mother squats, gives birth, and moves on; but a human mother is generally incapacitated and requires assistance. This biology is indeed so unintelligently designed that the author of Genesis 3:14 could only imagine it as a divine (if somewhat random) curse: Because Eve disobeyed and ate the fruit and now knows good from evil, says the Lord God, "I will greatly increase your pangs in childbearing; in pain you shall bring forth children."[3]

Eve's descendants, so to speak, the members of the genus *Homo*, whose adult brains will more than double in size over the course of about 2 million years, will solve that problem by tweaking the physiology of birth – for example, having the human baby rotate in the birth canal; and also by making birth social, rather than solitary, as in apes. Unlike an ape, a human giving birth generally has someone else around to help out. But this created another problem, for ape mothers are quite skittish about anyone who is a potential threat touching their infants, and do not tolerate it readily. To make birthing social, our ancestors would have to override or cancel this instinct, in order to be calm around grandparents, spouses, siblings, doctors, nurses, midwives, doulas, and whoever else, all taking an interest in the well-being of new mom and her baby.

The beginning of the genus *Homo* also roughly coincided with the adoption of a new tool: the hard, sharp edge. Even living chimpanzees understand the concept of jabbing a stick into a hole to get something: ants, termites, bushbabies, whatever. What is new is the development specifically of the first tool that archaeologists can identify as a tool millions of years later. To the student of paleolithic life, a stick is invisible in the archaeological record, and a bone wielded as a club – as in the movie *2001: A Space Odyssey* – is just a bone. But a sharpened rock is another matter entirely. Stone tools may be over 3 million years old, certainly over 2.5 million, and they get more complex and delicate as our ancestors' brains grow. Their discovery of cutting was as significant a technological breakthrough as their discovery of burning. And both of those came in handy because the brain requires a lot of energy to maintain, which in turn probably drove them to incorporate more meat in their diet than apes consume, thereby altering the microorganisms that populate the human gut, to help digest it.

The development of burning and cutting raises the question of when it may be inappropriate to burn or cut; we would hardly have survived if our remote ancestors had simply burned and cut things wantonly. After all, fire not only yields the possibility of cooking and cauterizing, but also the possibility of sacrificing and incinerating. Sharp stones can be used not only on carcasses and vines, but also on one another's throats. This begins a process that stretches to the present day: the coevolution of technologies and the lagging moralities that regulate them.

Australopithecus and *Paranthropus* are not known outside of Africa, but by 1.5 million years ago *Homo* was in southern Asia, and a bit later in southern Europe as well. These ancestors had smaller brains and larger jaws than we do, but their arms and legs were of human-like proportion and form. We don't know whether they were furry, naked, or clothed, or whether they built huts and could produce fire as needed. They were, however, successful – spreading across Africa, Asia, and Europe. Their success was due fundamentally to a new ecological relationship, at the base of the learning curve of human evolution, which increasingly involved three aspects. First, making usable things out of the available natural objects, and thereby constructing a physical environment consisting of raw materials. Second, transforming the environment itself into a more hospitable place than it was before they got there. Here the environment dialectically comprises both problems for people to solve and the raw materials for solving them. Humans consequently evolved not to passively occupy and adapt to an environmental niche, but to actively construct the environmental niches they inhabit. And third, creating a portable, imaginary environment as well – not simply tools, which you can take wherever you go, but also names, beliefs, stories, reputations, obligations, taboos, legends, ritual, courtesy, honor, dead ancestors, unborn descendants, and common sense, not to mention good old-fashioned knowledge – that is to say, culture. This latter mode of human adaptation and survival is universal, yet locally specific. All people have obligations and taboos, but the particular obligations and particular taboos vary by place and time. And although some taboos may have adaptive value, most are pretty ridiculous; they are all part of an imagined universe, a universe that doesn't correspond to ordinary reality – as do gazelles, trees, rabbits, pebbles, waterfalls, and newborn babies – but imparts order and coherence to reality by organizing, labeling, and regulating it. And, like tools, this imagined universe comes with you, and helps make your environment that much more uniform and familiar. Human evolution is consequently increasingly biocultural evolution, and we characterize the human condition itself increasingly as adaptable, rather than as adapted – since apparently any person can learn any language or lifeway, and humans don't appear to be particularly well suited biologically for much more than walking and talking.

The survival of the genus *Homo*, over the last million or so years, came in spite of the increasing difficulties in parturition, and were due principally to non-biological features. The expanding brain was clearly both a curse and a blessing, and opened up new venues for the construction of non-physical worlds with which to engage the physical world. The inferable biological changes that accompanied brain growth were the prolongation of youth (childhood), the prolongation of old age (menopause), and the inclination to trust others rather more than apes do (prosociality).

The prehistoric assistance in bearing and rearing big-headed babies that grow into immature children will come from several sources. One source is the new mother's own mother, for, unlike apes, who breed until they die, human females generally live considerably longer than they breed, which gives grandmothers an opportunity to assume roles as caregivers rather than as breeders. Another source might be as the result of an agreement with another family to provide a spouse. Moreover, the extended immaturity means that boys and girls grow up together, go through puberty, and yet are socially unready to work their own way into another group, as apes do. Unlike apes, then, adolescent human males need to consider certain females to be inappropriate sexual partners, and vice versa, and behave accordingly. And unlike the apes, brothers and sisters maintain lifelong bonds, which are disrupted in other species by sex-biased dispersal. Adolescent baboon males work their way into a different social group around age eight; adolescent chimpanzee females transfer into another social group around age eleven; but adolescent humans stick around for far longer. Siblings, mates, and grandmothers were always there to some extent as organisms, but in human evolution they come to assume significant social roles as well. Humans thus develop and elaborate three broad social relationships that are not part of the lives of apes: grandparents, spouses (creating in-laws and fathers), and post-adolescent opposite-sex siblings.

In these new social relationships that will construct the family in its myriad forms, we glimpse the origin of rule-governed behavior – prescriptive, in the obligations to a spouse; and proscriptive, in the control of sexual activity between siblings. In the grandmother, we glimpse the beginning of lineage, for a large brain is capable of imagining ancestors even before her. The moral and the social – what we might consider behavioral humanity – thus have a common root in the invention of family. We do not know when these ideas about kinship – and all the variations on it that a symbolling mind can create – arose, but they mark a significant departure from the life of apes over the course of the last few tens or perhaps hundreds of thousands of years.

We have a better grasp of physical humanity, which arose in the fossil record in Africa perhaps 200,000 years ago, manifested principally as a higher forehead and more rounded skull; a lighter body build; smaller face, jaws, and teeth; and a chin. These people eventually supplanted the other Ice Age peoples in the more northerly latitudes in Europe and Asia, who had larger, wider faces and brows, thicker skull bones, and lower, longer skulls. These peoples had been quite successful, having themselves spread over most of the world. Since their physical distinctions are often variable and subtle, anatomical evidence has long suggested genetic contact among these populations, and genomic data now reinforce this. Human populations trade and interbreed; and just because *we* make formal distinctions among these peoples primarily on the basis of their head shapes and genomes, there is no reason to think that they also did.

The material remains of human abstract thought are inferred from tools and zigzag designs upwards of 100,000 years ago, and carvings and cave paintings from about 35,000 years ago. While there is a heavy Eurocentric bias in the archaeological data, because Europe is where the bulk of the work has been done, very early art has been found in other parts of the world as well, notably in South Africa, the Near East, and Oceania. People were in Australia by about 50,000 years ago, and in America by about 15,000 years ago – and drawing wherever they went.

By the mid-twentieth century molecular biology had shown that the DNA base sequence, or genome, encodes the identifiable genetic variations we study. Evolution, understood as alterations to the genetic structure of populations, is consequently rooted in genetic change. But much of the difference and change that we experience is non-evolutionary – that is, not located in the DNA base sequences, which spread by chance or by the deterministic proliferation of favorable variants. There are two other general mechanisms for the production of differences among groups of people. The first is cultural change, the cognitive and behavioral bases by which some people, for example, combat one another with obsidian knives, others with machine guns, and others not at all – such differences are explained by the nongenetic processes of history, rather than biology. "Evolution" is often used metaphorically in these contexts, to refer to the history of society or technology, rather than biological history. The second general mechanism is epigenetic change, in which part of the DNA is altered in response to the conditions of life, but not the DNA base sequence itself, and these modifications, which affect the expression of genes but not their structure, are passed on. The relationships among evolutionary change, cultural change, and epigenetic change are the subject of much contemporary interest.

We are always quick to point out that humans are still evolving – but we mean that in a narrow sense. We don't seem to be evolving physically – our brains don't seem to be getting bigger, nor do our side toes appear to be shrinking. We do find that since food production, our jaws have been getting smaller and weaker, a small percentage of people have wisdom teeth that never erupt, and, in the last few generations, people are maturing earlier and getting larger; the etiology of these changes is unclear, but is probably epigenetic. Since evolution fundamentally involves changes to the gene pool, these physiological differences may not technically be evolution, but its facilitators. Nevertheless, human gene pools have indeed been clearly adapting to certain stressors, like malaria (via sickle-cell and many other genetic variants of the blood), altitude (among Tibetans), and perhaps even dietary milk (lactase persistence).

The Evolution of Inequality

If evolution is minimally changes to the gene pool, and many differences among species and among human populations are genetic, then what of the differences between the classes? Are the aristocrats and plutocrats different from the workers and homeless by virtue of microevolution, or for some other reason? After all, if the wealthy differ from the poor because of evolution, then this would imply that economic history is simply the unfolding of genetic destiny, and that there is an underlying naturalness to gross economic inequality: The rich essentially evolved to have better lives than the poor. In other words, it would imply that although inequality exists, it is not unjust, because the observable inequalities in human experience are expressions of a deeper, if invisible, natural order – as naive and intellectually compromised scientists from time to time have indeed vainly argued. Perhaps unsurprisingly, then, radical evolutionary theory and conservative politics have perennially found convergences, for here, the genome is imagined to contain natural barriers to social progress and equality.

Scholars across the rest of the political spectrum have instead tended to see social difference and biological/genetic difference as sometimes correlated, but rarely, if ever, causally related. Modern humans with different appearances, gene pools, and histories nevertheless seem to have more-or-less interchangeable brains; indeed, over the last few tens of thousands of years, the achievements of modern human groups have been dependent upon cooperation, motivation, and mobilization, not upon any representative individual's brainpower. Certainly the accomplishments of members or descendants of immigrant, colonized, or oppressed communities attest to this. In fact, the causal arrow connecting biology and the circumstances of life can even go in the opposite direction, with poverty and oppression inscribing themselves upon the body (and upon the DNA, epigenetically) by causing long-term stress, which affects the endocrine and immune system in subtle but significant ways.

The overextension of the concept of evolution to explain things like economic stratification is a confusion of categories that has significant biopolitical implications, and creates a science-and-society paradox for us. On the one hand, we are concerned that a segment of society does not take evolution seriously enough, and rejects it; hence this book. But on the other hand, we are also concerned about a segment of society that takes evolution too seriously, applying it extravagantly by invoking genetic processes inappropriately in human history. It is an abuse of evolution to invoke it as a rationalization of the differences between settlers and indigenes, exploiters and exploited, wealthy and poor, comfortable and afflicted. There is in fact nothing at all "natural" about billionaires in human biology or evolution. Of course, the fact that billionaires are biologically unnatural does not tell us how society should treat them. We make that moral decision ourselves.

This brings us to a place that many scientists are reluctant to visit: Morality. One of the key scientific lessons that gradually unfolded in the twentieth century was that facts and values are not separable from one another. An earlier generation of scientists had the luxury of maintaining that what they were engaged in was a world of "just facts," but chemists working on poison gas during World War I began to put the lie to that distinction, while doctors and physicists working on opposite sides of World War II buried it. We inhabit a world of right and wrong, not simply of true and false. In 1925, Clarence Darrow discovered that the textbook John Scopes had used to teach evolution – for which he was on trial in Tennessee (see Chapter 3) – also taught white supremacy and the need to sterilize the poor on account of their hereditary feeblemindedness. As a lawyer and defender of civil liberties, Darrow was astonished that the scientists did not seem to care about social justice as much as they cared about Darwinism. As soon as the trial was over, Darrow began attacking these same scientists – the ones whose views on extinct fish he had just finished defending – for their bad views, and their bad science, of living humans.

Eugenics, the idea that the poor constitute a genetic threat and so should have their breeding restricted, made Clarence Darrow uncomfortable, but it was being taught throughout the 1920s as part of the college curriculum in evolution and genetics. In fact, to oppose eugenics in America in the 1920s was to face the accusation of being anti-science, and even of being a creationist! Somehow it fell to the non-biologists to delineate a normative science of human evolution that was distinct and separable from a science of eugenics.

But that is a strange and unsustainable situation. It has to be up to the scientists themselves to recognize the social value in their work, and to face the consequences of having a poor moral compass. History shows us that modern political evils – whether racist, sexist, colonialist, or born of some other ideology that opposes equality – are impervious to progress in science. Where a biblical argument might rationalize the exploitation or subjugation of peoples because they were "created" to be unequal, a post-Darwinian could rationalize it in terms of people having "evolved" to be unequal. In either case, a political state of inequality is identified and explained by recourse to hidden differences in natural abilities or aptitudes, with the divine or genetic origin of those invisible innate gifts being largely irrelevant. The moral dilemma we face in the study of human evolution is when the legitimacy of science is appropriated for ignoble political ends; i.e., to perpetuate and exacerbate inequality and injustice. While racist evolutionary biology raises its head periodically, today the most normative abuse of evolution comes from "evolutionary psychology," which offers ostensibly Darwinian explanations of various aspects of modern socio-sexual life that most of the rest of us have long understood to be the products of historical, rather than naturalistic, forces.

One such argument is that "we" evolved to be naturally "mildly polygynous" – that is to say, we are genetically predisposed, on the basis of our sexual dimorphism in body size, to live in social networks where males have more than one female mate, but not vice versa.[4] Male baboons are much larger than female baboons and are socially polygynous, whereas male gibbons are the same size as female gibbons and are socially monogamous. By that yardstick, in our own species, since men are generally bigger than women, we are naturally a bit like baboons physically and thus presumably socially as well. This sounds plausible, as relating human behavior to its primate homologs ought to be, but it also encodes a familiar morality tale: a husband caught with another woman is merely obeying his urges, while a wife caught with another man is committing a crime against the natural order. And yet the merest acquaintance with primate biology shows that it cannot be that simple. Baboon sexual dimorphism is also expressed in the canine teeth, with male baboon canines being especially frightful. In the fairly monogamous gibbons, the female's canine teeth are just as frightful as the male's; by this metric then, humans are more like the monogamous (if philandering) gibbons, with equality of the canine teeth. Moreover, humans have evolved patterns of sexual dimorphism – such as the distribution of body fat, and the size of a bump behind the ear, known as the mastoid process of the skull – which have no apparent counterparts in the primates, and thus suggest different and non-comparable sexual evolutionary processes operating in our own ancestors.

What kinds of processes? Presumably symbolic – where prestige, a good name, wealth, honor, love, and parental consent become variables to be traded against the classic mating attractions of beauty or strength. But with patterns of sexual dimorphism that simultaneously indicate "natural" states of polygyny, monogamy, and non-comparability in relation to other primates, it certainly seems as though the evidence for our evolved socio-sexual system sums to zero. (And that's without even reasoning directly from lobsters to humans, as some popular modern psychology crudely does!)

Another such pseudo-evolutionary argument is sometimes made on the basis of experimental results from the 1940s, which showed that the fertility of a male fruit fly increased with the number of mates it had, while the fertility of a female fruit fly did not. Since the extra matings didn't help her reproductive output, why would she bother to seek them out? It would simply be a waste of energy. Therefore it makes evolutionary sense that the female fruit fly might be a bit more coy than the male. The conclusion drawn, however, was not about male and female fruit flies, but about males and females generally – who supposedly evolved to be transcendently randy and demure, respectively. Once again, though, some knowledge of the actual biology undermines the suggestion that this is what nature made us to be. A female fruit fly, after all, has a peculiar organ called a spermatheca, whose function is to store sperm to fertilize her eggs. In other words, she has a particular fruit fly specialization for not needing more than one copulation over the course of her fruit fly life. Some other insects share this physiology, but humans and our primate relatives don't. The male fruit fly, for his part, has sperm cells that oddly fail to undergo the near-universal genetic process of crossing-over,[5] which means that his sperm cells are not very different from one another, since crossing-over is one of the processes that scrambles the chromosomes every generation. So his sperm cells are far more alike than either their ovarian counterparts or their human counterparts are, since human sperm cells do indeed undergo crossing-over. If you're a male fruit fly, then, fertilizing eggs from multiple partners is an evolutionary strategy that will help offset the lack of genetic diversity among your sperm. But if you're a male human doing the same thing, you're just a jerk.

Which returns us to the question that troubled the civil rights lawyer Clarence Darrow back in the 1920s: Why should science justify jerkish behaviors and ideas – whether about controlling the reproduction of the poor, or the primitive brains of enslaved people, or the philanderer sowing his wild oats – by making them sound natural rather than bad? Who benefits when it does this? And why must it fall to non-biologists to open up the intellectual space within which one can be a Darwinian, but not quite so morally challenged?

This is not to say, of course, that modern creationists aren't racists or sexists, simply that they don't abuse the authority of science to validate their prejudices. That is because they don't possess the authority of science.

3
Attacking Evolution

One reason *The Origin of Species* is still readable over 150 years later is that Darwin assiduously avoided any mention of people, save for one fairly anodyne comment near the end: "Light will be thrown on the origin of man and his history." Finally, in the sixth and last edition of 1872, Darwin amended that sentence to read, "Much light ..."

The year before, Darwin had published his own ideas about humans, and, of course, suffused them with the prejudices of his age, class, and sex. Which is why *The Descent of Man* (1871) is such a tough read; you have to grimace every few pages, it seems. And indeed, it proved to be a vulnerable spot: a couple of generations later, the prominent creationist (and feminist) William Jennings Bryan taunted his interlocutors:

Darwin explains that man's mind became superior to woman's because, among our brute ancestors, the males fought for the females and thus strengthened their minds. If he had lived until now, he would not have felt it necessary to make so ridiculous an explanation, because woman's mind is not now believed to be inferior to man's.[1]

The theory of human evolution has clearly been value-laden since its very beginning. And while we often try to convince the public (and ourselves) that it is "just" about a descent from the apes, and adaptation by natural selection, it never actually is. Consider Ernst Haeckel, who believed that there were many species of living humans, at different distances from the apes; he was the leading expositor of first-generation Darwinism in Europe. Or Henry Fairfield Osborn, who not only believed there were many species of humans, but also understood evolution to dictate the involuntary sterilization of poor people; he made the cover of *Time Magazine* on December 31, 1928, as the leading expositor of evolutionary biology in America. Or the contemporary biologist Richard Dawkins, who links his understanding of evolution to his atheism; or Matt Ridley, who links his account of it to the free market and the non-renewable energy industry.

It should not be up to everyone else to disentangle the science from the politics; indeed it may be a fool's errand even to try. But that leaves a couple of important questions unanswered: What are the actual implications of evolution, and upon whom can we rely to elaborate them?

Germans Against Darwinism

Let us begin with the recognition that human evolution is a narrative tablet, upon which scientists inscribe stories of their ancestry, scientific stories co-written with cultural values. As *The Origin of Species* became well known, the most prominent European resisters against Darwinism within the scientific community were the first-generation German anthropologists, notably Rudolf Virchow and Adolf Bastian. Virchow was also arguably the leading biologist in Germany – a pioneer in social epidemiology and cellular pathology – a political liberal, and no more religiously devout than the next fellow. And yet he infamously rejected evolution, publicly stating in 1877 that "We cannot teach, we cannot designate it as a revelation of science, that man descends from the ape or from any other animal." He wouldn't accept Neanderthals as human ancestors (but then, of course, neither do we), nor would he accept "Java Man" in the 1890s. And yet, he wasn't exactly a creationist. He tried to clarify his position in 1884: "I never was hostile to Darwin, never said that Darwinism was a scientific impossibility. But when I pronounced my opinion on Darwinism [in 1877], I was convinced, as I still am, that the development which it had taken in Germany was extreme and arbitrary."[2]

Virchow, it seems, was not put off so much by natural selection and the descent of people from apes, but rather by a particular expression of those ideas – specifically that being put forward by Ernst Haeckel. Haeckel not only saw evolution as a rise from the lowly ameba to the Teutonic military state, but also saw non-Europeans as zoological intermediates between Europeans and the apes, and as different species altogether. Virchow and his friend Adolf Bastian, on the other hand, founded the Berlin Anthropological Society in 1869 on the idea of "the psychic unity of man" – the assumption being that, to effectively compare the diverse lifeways within the human species, the brain (or the genes, or race, or biology) must be a constant. Since all people are built more or less the same way, the relevant human differences must be acquired, not innate, and the relevant processes that made those differences are historical, not evolutionary. This, it hardly needs noting, was the opposite of assuming that humans comprised a dozen distinct species. If, to European audiences, evolution meant that different peoples were different species, then a science founded on the idea that they all possess a "psychic unity" would have to oppose that particular theory of human evolution.

After Virchow's death in 1902 there was effectively no opposition in Germany to evolution. But the Darwinists' victory came at a considerable price. When the Europeans were fighting a Great War a few years later, and before America entered to make it a World War, *The New York Times* published an essay by Ernst Haeckel himself, rationalizing the German Empire's militarism by recourse to evolution. "The beneficial progress of evolution," he explained, "outweighs the injurious effects of the regressive development during the war."[3] Then biologist Vernon Kellogg published a bestseller called *Headquarters Nights*, which recounted conversations he had as a neutral observer visiting the German high command. According to Kellogg, the German officers were well-educated but had a strange biopolitical idea of what they were fighting for; namely, "a point of view that justifies itself by a whole-hearted acceptance of the worst of Neo-Darwinism." Apparently, they saw "mankind as a congeries of different, mutually irreconcilable kinds, like the different ant species," and in consequence, that human group which is in the most advanced evolution-ary stage as regards internal organization and form of social relationship is best, and should, for the sake of the species, be preserved at the expense of the less advanced, the less effective. It should win in the struggle for existence, and this struggle should occur precisely that the various types may be tested, and the best not only preserved, but put in position to impose its kind of social organization . . . on the others, or, alternatively to destroy and replace them.[4]

Kellogg was distancing himself intellectually from the evolutionary theory he was describing, but he was a day late and a dollar short, for Haeckel's Darwinism was clearly the dominant view in Europe and had already influenced global geopolitics. Nor was this "social Darwinism," a label applied much later to ideas in mostly Anglophone intellectual circles, rationalizing the exploitation of labor by the nineteenth century "robber barons." Haeckel's idea was more specifically about the biological imperative, and ultimate evolutionary benefit, of perpetual and merciless war.

Now, suppose that you were a pacifist, you felt war to be morally repugnant, and you could see that while evo-lution may be multivocal, one strong message it sends out to millions of people worldwide is that war is good. How might you come to feel about Darwinism?

The Scopes Trial

One such moralist was William Jennings Bryan – former Secretary of State, three-time presidential candidate, fem-inist, Prohibitionist, pacifist, isolationist, writer, orator, and Christian – who by the early 1920s had become convinced of the moral depravity of the theory of evolution. And not because he was poorly read; precisely the opposite, because he was well-read, and knew about the Darwinism that rationalized murder and avarice. To Bryan, Darwinism was a "menace to fundamental morality" and thus "harmful as well as groundless." Of course Bryan was not a competent judge of its scientific grounding, but he could – indeed, any thoughtful Christian could – readily see Darwinism's harm.

Bryan became an outspoken creationist toward the end of his life, and when Dayton, Tennessee became the national focus in 1925 for charging schoolteacher John Scopes with the crime of teaching evolution, Bryan quickly volunteered to be guest prosecutor. He began by appearing to deny that we are mammals, in order to protect the children:

And then we have mammals, 3500 [species]. And there is a little circle and man is in the circle. Find him; find man. There is the book that they were teaching your children, teaching that man is a mammal and so indistinguishable from other mammals that they leave him there with other mammals, including elephants. Talk about putting Daniel in the lion's den! How dare these scientists put man in a little ring like that with lions and tigers and everything that is bad? . . .

In the notebook, children are to copy this diagram and . . . show their parents that you cannot find man. That is the great game to put in the public schools, to find man among the mammals, if you can . . .

Shall we be detached from the throne of God and be compelled to link our ancestors with the jungle? Tell that to the children![5]

Now let's pause right there. Actually, there is a subtle case to be made for challenging the classification of humans within the Class Mammalia, but Bryan isn't making it. After all, the Bible does classify animals (in Leviticus 11 and again in Deuteronomy 14), and there is no category corresponding to mammals. The Bible strives to impose order on the animals, as science does, but it does so for different reasons, and with different criteria. The ancient Hebrews classified animals not by their overall similarity or common ancestry, but by two sets of key features: first, by where they lived (air, land, water); and second, by their locomotor anatomy (flying, jointed legs, swarming, crawling, etc.). The purpose was explicitly to divide the biome into clean and unclean animals, and thereby specify what could be eaten. But there is no category corresponding to the warm-blooded, hairy, lactating vertebrates. What we call mammals, the Bible actually classifies in several different places – in the waters, without fins and scales; in the air; on the ground, walking; and on the ground, swarming. Arguing from the standpoint of the Bible, then, one could argue that we are not mammals, for the Bible acknowledges no such category. Classifying as scientists do, we are mammals, but classifying any other way, we might well not be. So perhaps the fact that the scientists classify organisms by patterns of similarity, taken to reflect common ancestry, does not necessarily mean that everybody else has to classify things using their criteria.

But Bryan didn't make that argument. He just wanted, as Benjamin Disraeli had wanted a few decades earlier, to be with the angels.

On the opposite side from Bryan, the greatest trial lawyer in the land, Clarence Darrow, was working *pro bono* for the only time in his career. As the trial wound down, Darrow pulled a surprise and called William Jennings Bryan himself to the stand as a biblical expert, grilling him relentlessly on ancient history, comparative religion, and the Bible. The most interesting moment, however, came when Bryan volunteered that he wasn't even a young-earth creationist. He believed that the days of creation were not precisely 24 hours long, and said so quite casually. And quite surprisingly. Based on the "begats,"[6] an Irish bishop named James Ussher had calculated back in the seventeenth century that the universe's first day came just a few thousand years ago; indeed, the day was October 23, 4004 BC. It was even printed in the margins of the Bible that was used to swear the witnesses in. But when Darrow asked Bryan if he thought the earth was only a few thousand years old, Bryan quickly responded, "Oh no, I think it is much older than that." Darrow prodded him again, "Do you think the Earth was made in six days?" "Not six days of twenty-four hours," he answered. Both sides were stunned, for they had assumed Bryan was a young-earth creationist, following the writings of an obscure Canadian anti-geologist named George McReady Price.

Darrow famously won the argument, leaving Bryan spluttering that "this man, who does not believe in a God, is trying to use a court in Tennessee to slur at it, and while it requires time, I will take it!" And Darrow, who always got the last word, got the last word, "I am examining you on your fool ideas that no intelligent Christian on Earth believes!"

But Darrow lost the case. Although the decision would be reversed in a higher court in 1927, Scopes was indeed found guilty of violating the law. The Scopes trial, however, had two interesting codas. First, five days later, William Jennings Bryan was pronounced dead. This is interesting theologically, for the simple reason that it had absolutely no theological consequences. Suppose, however – as the biologist and philosopher Julian Huxley observed – it had been Clarence Darrow who died straight after the trial. That would have been loaded with theological significance, wouldn't it? The theology is asymmetrical: the people who look for such implications would have seen the death of their opponents' champion as a divine punishment; but from the death of their own champion, they inferred no such heavenly message.

The second coda was a false memory of the trial, fictionalized in a 1955 Broadway play and 1960 movie called *Inherit the Wind*. The dramatization used the Scopes trial in the 1920s as a backdrop to criticize the contemporary suppressive politics of the McCarthy era. Although the play used some actual courtroom dialogue, most of it was composed, and the stage characters were necessarily less nuanced than the actual people they were based on. Yet even today, many people misremember the Scopes trial through its reconstitution in *Inherit the Wind*.

Science Has Its Story, and I've Got Mine

Meanwhile, the anti-evolution statute remained on the books until the late 1960s, when a Supreme Court decision eventually rendered all such laws unconstitutional. Creationism did not disappear, but merely licked its wounds; indeed, there is always plenty of anti-intellectualism to go around in American life. One particularly weird form of it arose in 1950. On June 18 that year, the top nonfiction bestseller in *The New York Times* was *Worlds in Collision*, by the psychoanalyst Immanuel Velikovsky.

Velikovsky was a modernist and evolutionist, but shared a critical assumption with the creationists – namely, that the Bible (specifically Exodus in his case) is "really" true. For Velikovsky, however, this did not mean literally true – and here he borrowed from the worst of nineteenth-century biblical rationalism. To Velikovsky, the Bible is not a historical chronicle, but it does record historical events, albeit misunderstood, misreported, and misrepresented over the centuries. So, he reasoned, given that the authors of Exodus were writing about something in history, but that miracles don't happen, what was it that they were remembering so poorly as the "miracles" of the Ten Plagues of Egypt and the Parting of the Red Sea?

Velikovsky reasoned that, after the Nile had "turned to blood" (Plague 1), the frogs found it uninhabitable and died (Plague 2), their decaying corpses sustained insects (Plagues 3 and 4), and disease eventually found its way up the food chain to cattle and people, while unusual meteorological phenomena continued (Plagues 5 through 10). But what could have set off this chain of events by turning the Nile to blood? Or by being at least mistaken for blood by the Semitic Bronze Age tribes who were experiencing these unfolding tragedies?

Herein lay Velikovsky's originality. What they mistook for blood was actually red dust falling from the sky onto the Nile delta from the surface of the planet Venus, which had recently become a planet after having been blown out of the Great Red Spot of Jupiter as a transient comet. A bit later its gravitational field would part the Red Sea, and with help of the planet Mars and some trumpets, would bring down the walls of Jericho. If that sounds like a lot to digest, it sounded that way to the astronomy community as well, even before *Worlds in Collision* hit the bestseller lists in 1950. Astronomy, after all, was the only thing standing in the way of Velikovsky's interpretation of Exodus. And the reaction from astronomers was swift and furious: He is wrong! His story about the solar system is terribly inaccurate!

The battle, however, should not have been over whose story was more accurate, but over how scientific stories get made. Velikovsky had no criteria for distinguishing between biblical passages to be understood as mythology or exaggeration or allegory, and those to be understood as historical fact. More significant, however, is Velikovsky's prioritizing, by which astronomy gets thrown under the bus in order to make his idiosyncratic interpretation of the Bible story work at all. And of course his version is not even the Bible story itself, but a weird retelling of it, in which nobody can tell blood from rusty water, sticks don't turn into snakes but the Red Sea really does part, and nobody notices that everyone is dropping dead (not just first-born Egyptians). Here the biblical text is neither chronicle nor fable, but simply a set of guidelines for creatively composing your own historical narrative.

So for that we are to sacrifice the entire science of astronomy? Yes indeed, said Velikovsky. You've got your story about the solar system and I've got mine.

The scientific community did not take this lightly, and battered Velikovsky. Nevertheless, he had shown that people would read and like a book about science by an iconoclastic author who knew next to nothing about the relevant material, but was willing to challenge and flat-out reject the science, and to make up his own counter-science to explain a miraculous story in the Bible.

We don't encounter Velikovsky much anymore, except on the internet. But his approach – validating a biblical story by overriding the relevant science and simply declaring it to be false – created a template for the next phase of creationism, as well as for the subsequent anti-science fad of "ancient astronauts."[7]

A decade after Velikovsky's anti-astronomy biblically rooted story, creationists produced their own anti-geology biblically rooted story. Rather than criminalizing evolution, they would instead begin simply by rejecting the science of the scientists and making up their own. Their precedent came from a Seventh-Day Adventist sourcebook from the early twentieth century, which was effectively repackaged in the 1961 book *The Genesis Flood*, by John Whitcomb and Henry Morris. Here, however, the creationism was that of the literal, young-earth six-day creation, a position that had been – and there is no polite way to say this – too ridiculous for even William Jennings Bryan to hold forty years earlier. Whitcomb and Morris, uncredentialed in the relevant geology, merely rejected that science and proposed instead that all geological features everywhere were the recent consequences of Noah's flood. By thus dismissing the geological science that had emerged a generation or two before *The Origin of Species*, Whitcomb and Morris actually only attacked Darwinism tangentially.

Now young-earth creationism, which had been out of favor even in creationist circles since the Scopes trial, began to make a comeback in evangelical Protestant churches. By the late 1970s, a new young-earth "scientific creationism" was promoting itself as a modern, scholarly alternative to evolution. And since its defenders imagined biblical literalist creationism to be the only alternative to evolution, they considered that any difficulty for evolutionary theory was evidence in favor of their young-earth creationism.

Both the evidence and the logic were scant and weird. Dinosaur and seemingly human footprints were apparently seen together in Texas, which proved that humans and dinosaurs coexisted in the Garden of Eden, and therefore by implication that all of biology and geology is wrong, the Bible is literally true, and humans are not descended from apes. The second law of thermodynamics, the one about entropy increasing in a closed system, proved that evolution, which generates complexity, is impossible, so the Bible is literally true, and therefore humans are not descended from apes. Anomalous properties of the element polonium embedded in some igneous rocks proved that the earth is only a few thousand years old, which means that the rest of science is wrong and the Bible is literally true, and therefore humans are not descended from apes. It is hard to imagine the defense mechanism of the bombardier beetle evolving, and therefore it could not have evolved, and therefore *did* not evolve (Richard Dawkins has called this "the argument from personal incredulity"[8]), and therefore nothing evolved, the Bible is literally true and humans are once again not descended from apes.

If the creationists couldn't make Darwinism illegal, they could at least mandate that their alt-scientific ideas be introduced in science classes alongside Darwinism, as a reasonable (perhaps an even *more* reasonable) alternative. Eventually, the state of Arkansas was sued over its "creation science" law, and lost at the Federal Court level.[9]

The 1981–82 court case, *McLean vs. the Arkansas Board of Education*, was culturally significant enough that Judge William R. Overton's decision was published in full in the journal *Science*. Overton's ruling established that "creation science" was not science, but a different set of practices with "no scientific merit or educational value as science." "The creationists' methods do not take data, weigh it against the opposing scientific data, and thereafter reach the conclusions . . . Instead, they take the literal wording of the Book of Genesis and attempt to find scientific support for it."[10] Moreover, added the judge, there was something distastefully duplicitous in the way the creationists represented their position. The creationist law isn't really about science, he wrote, or the general advancement of knowledge at all; it "is a religious crusade, coupled with a desire to conceal this fact."

This highlights an interesting paradox. By 1982, there had been a lot of intellectual movement over the many decades since the recognition of our descent from the apes. Theology had progressed, beginning to engage with evolution in creative ways. Philosopher-theologians Teilhard de Chardin, Alfred North Whitehead, and Hans Jonas (a Catholic, a Protestant, and a Jew, respectively, although they never walked into a bar together) all sought meaning in the science, and a new generation of theologians was being influenced by their writings. Our understanding of evolution had progressed as well. At the time of the Scopes trial in the 1920s, it was known crudely that the bloods of humans and apes were more similar to one another than the bloods of horses and donkeys were to one another. But by the 1980s, scientific research had established that genes build an animal, that genes are made of DNA, that changing the DNA changes the animal, that changing the proportion of DNA variations in a species changes the average appearance of the species, and that the closer we look at the genes, the more similar human and ape appear to be. In contrast to both theology and biology, creationism had actually regressed, adopting the most radical and paranoid model that saw science as an enemy, rather than as an intellectual ally, in the general quest to understand and cope with life.

Intelligent Design: Old Wine in New Bottles[11]

Thwarted by their attempts to criminalize evolution, or to balance biblical literalism alongside it, creationists at the end of the twentieth century adopted a third legal strategy to undermine evolution, and to pretend that they had something new: Intelligent Design. Intelligent Design makes no explicit claims about the age of the earth, thus attempting to embrace both old-earth and young-earth creationists; neither does it make explicit claims about the attributes of the "intelligent designer," thus attempting to embrace people of diverse faiths, who may be united simply in their hope that life has purpose and meaning. Consequently, Intelligent Design is not so much a theory as an anti-theory: However old the earth and its species may be, and whoever and however they came to be, nevertheless evolution is definitely wrong and we are definitely not descended from apes.

Attacking Evolution

The general idea that human evolution had a little supernatural help along the way is impossible to falsify, for that supposed help lies outside the material universe in which falsification is possible. Nevertheless, the idea is certainly compatible with the descent of humans from apes over the last few millions of years, and has even been embraced in some form by some idiosyncratic scientists from time to time.

Notable among those idiosyncratic scientists was the co-founder of Darwinism, Alfred Russel Wallace. By the late 1860s, Wallace had become a devotee of spiritualism, including charlatanry of various kinds: séances, clairvoyance, animal magnetism, apparitions, table-rapping, telepathy, spirit photography, and the like. In the 1860s he was content to simply imagine human evolution being aided a little by spiritual forces, but he became more deeply involved as he got older, telling *The New York Times* decades later, "I imagine that the universe is peopled with spirits – that is, with intelligent beings – with powers and duties akin to our own, but vaster."[12]

Likewise, the Scottish-South African paleontologist Robert Broom did not doubt "that man has been descended from an anthropoid ape," but nevertheless speculated freely and quaintly about how it happened. "From the fact that beauty and cruelty are rarely found together, we seem driven to the conclusion that there are various spiritual forces behind evolution, and that the agencies that have loved beauty are not the same agencies as those that have evolved the birds of prey and the venomous reptiles and poisonous insects." Not only were the benign forces helping the hummingbirds and the malign forces helping the tarantulas, but they all had a bigger plan – to help us, for "the production of man has been the chief purpose of it all."[13]

Neither scientist was fully comfortable with the materialist basis of scientific knowledge, and both considered their spirituality to be complementary to their evolutionary biology, not an alternative to it. And that is where they part ways with contemporary Intelligent Design. Now, whatever spiritual forces may be out there creating the history of life are invoked not to augment Darwinism, but to contradict it.

Intelligent Design's ostensible target is natural selection, the idea proposed by Charles Darwin to explain how species could change adaptively. The Intelligent Design argument, in its bare bones, is simply that if natural selection does not explain how species change adaptively – and its defenders are at great pains to assert that it does not – then it must be Jesus. Of course, they don't like to name the Intelligent Designer publicly, because they try to conceal their basis in fundamentalist Protestant biblical literalism. In fact, though, much of their rhetoric and support base was imported directly from the earlier "creation science." When Intelligent Design arrived in federal court in 2005 as *Kitzmiller vs. the Dover (Pennsylvania) Area School District*, the judge ruled that not only was it an attempt to get a religious ideology into a science class, but it was again accompanied by a layer of deceit: for example, an Intelligent Design textbook had repackaged "creation science" as "intelligent design" by merely substituting the latter phrase for the former in the text. Pretty clearly, its promoters didn't want you to know what they were really up to. In fact, their own correspondence showed that they saw Intelligent Design as a "wedge" by which to introduce conservative Protestantism into the public curriculum. Its goal, said the judge, "is not to encourage critical thought, but to foment a revolution which would supplant evolutionary theory with [Intelligent Design]."

Despite its defeat in the courts, Intelligent Design gained some traction outside of conservative evangelical Protestant circles, usually from people who didn't realize that they were being swept up in a radical literalist theology. That was, in large measure, because the promoters of Intelligent Design are far clearer about what they don't believe (evolution) than about what they do believe. However, the very idea that God's existence requires proof from the natural world is alien to the Bible. The ancient Jews took the existence of God on faith, and were content to admire the properties of the universe and its contents; it was the pagan Greeks of the ancient world who felt the need to prove God using nature. These ideas – that nature and reason can be employed to augment faith – entered mainstream Christian thought much later, through the influence of the work of Augustine and Thomas Aquinas. Consequently, the arguments were familiar by the time William Paley codified them in his *Natural Theology* (1802); arguments that the philosopher David Hume had already demolished a generation earlier in his *Dialogues Concerning Natural Religion* (1779).

It was Paley who introduced his readers to the idea of the watch whose existence implies a wise watchmaker, inviting them to see the human eye as the product of a wise eye-maker. His book was read by Darwin in college, and the argument has not evolved much since then. Today, Intelligent Design recasts it in relation to

the "irreducible complexity" of cellular processes, which supposedly implies the existence of a wise cell-maker.

But is that inference valid? What are the attributes that are so complex that they necessitate an eye-maker or a cell-maker? Consider the state of Texas. Its shape is very distinctive, complicated, and easily recognizable. How do you suppose it got that way? One possibility might be through a series of historical processes – complex, but ultimately knowable – such as the geological action of the Rio Grande, Sabine, and Red rivers; the coastline of the Gulf of Mexico; various wars and treaties with Indians, Mexicans, and French; and agreements worked out with other states. Another possibility might be that it was directly fashioned by the hand of a divine state-maker. If the former seems more likely, or at least frames the question as answerable, if difficult, the same is true of Paley's watch. After all, just as the state of Texas has a history that explains how it came to look as it does, so does the watch; and wise watchmakers can only ply their trade thanks to the many watches they made that didn't work very well, and to the trial-and-error successes and failures of prior generations of watchmakers. But that's not how we think of God's creation. As Hume put it back in the eighteenth century:

And what surprise we must feel when we find him a stupid mechanic, who imitated others, and copied an art, which, through a long succession of ages, after multiplied trials, mistakes, corrections, deliberations, and controversies, had been gradually improving? Many worlds might have been botched and bungled, throughout an eternity, ere this system was struck out; much labor lost; many fruitless trials made; and a slow, but continued improvement carried on during infinite ages in the art of world-making.[14]

In other words, the natural order is not really like a mechanical contrivance at all, for the latter has a learning curve – a history of equally or less-intelligently designed objects – while the natural order doesn't seem to. And that is a really significant difference for the purposes of this analogy. Things are created historically. Thus to understand their creation, we must understand their history. Scholars realized this hundreds of years ago.

The analogy between the products of nature and the products of human contrivance is thus powerful, but ultimately terribly misleading. Consider the obvious fact that Seikos, Rolexes, Citizens, and Timexes all have different designers. Is Intelligent Design a theory of polytheism then? Or is nature not really so much like a watch after all?[15]

Nor is it clear exactly what particular properties necessitate the inference of the direct hand of God. Snowflakes are remarkably intricate, but nobody seriously thinks that God fashions each one individually. Sure, one can see God as the maker of the laws governing crystal formation, but such a theology – with nature as an immediate cause, and God as an ultimate or primary cause – is at least as compatible with evolution as with biblical literalism! We simply don't need to infer the direct action of God to explain the intricacy in the world.

In addition to intricacy, we can see order without the direct action of God as well. No matter how hard I try to make my Italian salad dressing disordered by shaking it up, the earth's gravitational field acting on materials of different densities always sorts it out with the oil on top, vinegar in the middle, and herbs on the bottom. We don't really need to think that God is actively pushing the herbs down and the oil up. Again, we just don't need the direct intervention of God to explain the order in the world. Of course, God may have established the laws of gravity, as scholars since Isaac Newton have speculated. Theologically, we are not necessarily calling the existence of God into question, but the object and manner of God's creative action.

And finally, can we identify functional complexity without invoking the hand of a heavenly complexity-maker? Indeed we can, argued the economist Adam Smith back in the eighteenth century. We don't need rules and regulations – design – to construct a working economy, he argued. We just need individuals pursuing their own best interests, and the system will run on its own, in complex and beneficial ways, as if guided by an "invisible hand" – Smith's famous metaphor in the founding document of free-market capitalism, *The Wealth of Nations*. A functioning and efficient system can indeed arise spontaneously (if you have any faith in capitalism).

So we can see form, intricacy, order, and complexity where it would be ludicrous to invoke the direct action of God. Does this mean "intelligent design" is wrong? No, but it means it is superfluous. Just as Newton showed that gravity could be understood as a natural phenomenon, Darwin showed that adaptation can be understood as a natural phenomenon. But that doesn't address the question of whether they *ought to be* understood as natural phenomena. The short answer to that question is "yes," but only because that is an assumption of science. In science, miracles are cop-outs; we don't know that nature is all there is, but we do believe that natural phenomena are best explained by natural processes. The superiority of the germ theory of disease over the evil spirits theory of disease is well attested, although there are many illnesses (most notably, mental) that the germ theory doesn't help us with. Yet we still look within the natural order for explanations, because, well, we're scientists.

If we can find form, intricacy, order, and functional complexity without imagining the hand of God busily at work producing them, then science is best served by examining the processes that do indeed produce them. That is why it sounds so theologically clumsy today to argue that a cell, or its flagellum, must not be the product of history, but of spiritual handicraft. The universe and all it contains, it seems, are not so much like 200-year-old timepieces after all.

In fact, the analogy breaks down even further when you consider that William Paley made the original comparison before watches were mass-produced. If today's watches indicate anything about the creative process, it's that watches are not handcrafted, but precision-generated. That is to say, the intelligent designer actively designs the watch factory, but not the individual watch – which, for the theist, would seem to provide a better argument by analogy for God's role as an ultimate or primary cause, the designer of evolving systems, one of which secondarily produced humans from an ape lineage.

4
Biblical Literalism and Rationalism

What have scholars known for the last 200 years that modern biblical literalists or inerrantists don't know? It is that, first and foremost, biblical literalism is self-contradictory. Consequently the literalist can never actually follow "the Word of God" since the "Word" itself is multivocal. If you expect the Bible to speak with a single voice, you will be surprised when you actually read it carefully. So if you present the Bible today as univocal, you are actively misrepresenting it, which ought to be a sin of some sort. Here are a few simple and relevant examples.

> So God created humankind in his image [on Day 6], in the image of God He created them, male and female he created them. (Genesis 1:26)

This certainly makes it sound as if God created man and woman simultaneously, which would presumably be the take-home gender lesson of creation. After creating plants on Day 3, and animals on Days 4 and 5, God gets around to people on Day 6, and makes men and women, together. And that is the way we would probably see it, if it were not for the subsequent chapter, which tells a somewhat different story. In Genesis 2 the creative actor is called Lord God, and He makes man first, then the plants and animals, and then woman.

> In the day that the LORD God made the earth and the heavens, when no plant of the field was yet in the earth and no herb of the field had yet sprung up . . . then the LORD God formed man from the dust of the ground, and breathed into his nostrils the breath of life; and the man became a living being . . . [O]ut of the ground the LORD God formed every animal of the field and every bird of the air . . . So the LORD God caused a deep sleep to fall upon the man, and he slept . . . And the rib that the LORD God had taken from the man he made into a woman. (Genesis 2:4–22)

Considering that all this creating has already been proceeding on Days 3, 4, 5, and 6, it isn't even clear just which "day" is being summarized here. After all, we had just read that people were created *after* plants and animals. Now we're reading that man was created, *then* plants and animals, *then* woman. These stories are different, incompatible, and are labelled as such in the text, with one story being a creation by "God" and the other by "Lord God." For hundreds of years it never mattered, because no serious scholars thought that it was necessarily describing a single material, historical act. Sure it *might* have been describing something literal, but there was no reason to think that it *had* to be; what scholars thought was important was to figure out what the text means – which is usually not obvious, or it wouldn't be taking scholars to figure it out. But the modern biblical literalist must take one of those two stories and pretend the other doesn't exist, or else do some fancy editing. Usually they go with the first story, and replace Day 6 with most of the second story. That creates coherence, but it is achieved by extensive hermeneutics, the very opposite of a simple literal reading.

Another familiar story also comes in two forms, which self-identifying biblical literalists ignore or suppress. In Genesis 6:19–20, God tells Noah, "of every living thing, of all flesh, you shall bring two of every kind into the ark, to keep them alive with you; they shall be male and female . . . two of every kind shall come in to you, to keep them alive." Once again, it sounds familiar and straightforward: Noah takes two of everything, on orders from "God." But once again, immediately afterward, "the Lord" comes in and gives Noah slightly different instructions.

> Then the Lord said to Noah, "...Take with you seven pairs of all clean animals, the male and its mate; and a pair of the animals that are not clean, the male and its mate; and seven pairs of the birds of the air also, male and female, to keep their kind alive on the face of all the earth." (Genesis 7:1–3)

Uh-oh, the ark suddenly just got a lot more crowded. There are now twelve more representatives of a heck of a lot of species for Noah to load. Not two oxen, but fourteen; not two sheep, but fourteen; not two goats, but fourteen. Also fourteen deer, gazelles, roebuck, wild goats, ibexes, antelopes, mountain sheep, and anything else that the Hebrews considered "clean." Not only that, but Noah's aviary suddenly got huge: only two eagles, vultures, ospreys, buzzards, kites, ravens, ostriches, nighthawks, sea gulls, hawks, little owls, cormorants, great owls, water hens, desert owls, carrion vultures, storks, and herons – according to the list of dirty bird-ies given in Leviticus 11 – and fourteen of *every other species*.

Once again, a literalist has to take one description of the ark's menagerie and ignore the other. That is either naive or disingenuous, and in neither case does it honor the sacred text. The sacred text is multivocal. Someone who takes the Bible seriously, but not literally, has no difficulty with these contradictions, for they are simple consequences of many centuries of composing and compiling a complexly structured literary corpus of the ancient Hebrew and Christian worlds.

The inaccuracies and inconsistencies in the Bible were not considered particularly important by Jews and Christians for many centuries. The very fact that early Christians canonized four different versions of osten-sibly the same story shows that clearly enough. Only Matthew's gospel contains a Sermon on the Mount. Only John's gospel lists a disciple named Nathaniel. Only Luke associates the birth of Jesus with a Roman census that took place in 6 A.D. John's gospel even says that Jesus died a day earlier (Passover preparation day) than the other gospels (Passover day). These apparent inconsistencies have bothered Christians occasionally – such as a second-century scholar named Tatian – but not much. These aren't problems for a pious Christian; only for a literalist.

The ancients knew better than to take their sacred texts at face value.[1] The early Church father Origen, who helped to assemble the Christian canon, didn't take it literally. Neither did St. Augustine, two centuries later. Christian theologians, from the very beginning, have understood Scripture as requiring appropriate interpretation to be understood properly. How could it be otherwise? To take something seriously, you must understand it, and understanding it means the opposite of taking it literally. Taking it historically is what you should be doing, if you want to understand it – learning what it's composed of, who composed it, and when, and how. That might help you understand the meaning of the words. The King James Bible, after all, mentions unicorns in several places – and however sacred you may hold that venerable tome, you probably don't want your theology to hang on the existence of unicorns.[2]

Biblical Bibliographies, Geometry, and Figures of Speech

Now suppose you were a teacher, charged with educating children about astronomy, and one of your pupils says something that you hadn't heard before, and which sounds a bit weird. So you ask them for their source, and they tell you that it's from Carl Sagan's well-known book, *Cosmic Children of the Milky Way*. That might sound convincing until you realize that there is no such book, neither by Carl Sagan nor by any other reputable astronomer. Then you might reject the pupil's point entirely, because you know that if the kid had to make up the citation for something that probably isn't true in the first place, then there probably isn't a real reference – it was just something the kid made up and then made up where it came from. That would be a fairly reasonable deduction; for the act of citing a non-existent reference impugns one's credibility. You can't just cite a non-existent reference and expect people to believe you.

Then you go home and read your Bible. Specifically, Joshua 10:12–13, which says,

> On the day when the Lord gave the Amorites over to the Israelites, Joshua spoke to the Lord; and he said in the sight of Israel, "Sun, stand still at Gibeon, and Moon, in the valley of Aijalon." And the sun stood still, and the moon stopped, until the nation took vengeance on their enemies. Is this not written in the Book of Jashar? The sun stopped in midheaven, and did not hurry to set for about a whole day.

You say to yourself, "Gee, if the Book of Jashar says so, then it must be true." And you read on, because you can't wait to get to the Book of Jashar, because it sounds so good. Your anticipation is piqued even further when you get to Second Samuel 1:18, which says, "He ordered that The Song of the Bow be taught to the people of Judah; it is written in the Book of Jashar."

So you keep reading, through First and Second Chronicles, through the gospels, through the epistles, and through Revelation. And you finally realize the awful truth. *There is no Book of Jashar.* It's as real as J. K. Rowling's *History of the Weimar Republic.*

But that's okay, because you know about the Bible. You know that it is a collection of sacred writings of diverse kinds, by different writers at different times. You know that it was culled from a broader set of sacred writings, some of which (like "Bel and the Dragon") are still held sacred by some (Catholics) and not by others (Jews and Protestants); some of which (like "Jubilees") were held sacred by sects no longer with us (the Dead Sea Scrolls community) and still are by less familiar modern churches (like the Ethiopian Orthodox); and some of which (like "Jashar") maybe weren't quite so sacred, since they managed to get lost over the ages.

But suppose you are a biblical literalist, and believe that the Bible is a single, self-contained sacred book that needs to be taken at its word. How could you possibly make sense of the fact that the Word of God cites a non-existent source? It must mean that the canon you venerate is, at the very least, incomplete. But once you acknowledge that, you leave yourself vulnerable to the possibility that the Bible has also lost the Book of Evolution along with the Book of Jashar, so why not just accept the science?

The issue at hand is theological, not scientific. How do we understand the Bible? Can it be sacred and insightful, the Word of God, while at the same time also being metaphorical, allegorical, mythological, poetical, and compiled? If not, then you have a big problem with the two little passages about the Book of Jashar. It means that your god's "Word" would not meet the standards of a mortal undergraduate's term paper, for it contains invalid citations.

And the Book of Jashar is only the beginning.

What would you think if you were reading a book, and you found that two chapters in different places were the same? Is the author having a sly joke? Did the printer make an error? It once happened to me – I bought a book in Nairobi, and it turned out to have two copies of Chapter 8, and no Chapter 9. Printer error. Fortunately, a helpful colleague later xeroxed the missing chapter for me. But in your Bible, Second Kings, Chapter 19 is the same as Isaiah, Chapter 37. Darn near word for word. This is a problem for the literalist, but not for the rationalist, who sees the text as a long-term literary product,

composed, revised, edited, and redacted. And occasionally redundant. And occasionally redundant.

Consider this: Second Chronicles 4:2 and First Kings
7:23 describe a part of the temple being built by King
Solomon. "Then [Solomon] made the molten sea; it was
round, ten cubits from rim to rim, and five cubits high.
A line of thirty cubits would encircle it completely."
But if the object in question was circular and 10 cubits
in diameter, would the circumference actually be 30
cubits? Of course not. Circumference is diameter times
π, and π is 3.14, which means that the circumference
would be 31.4 cubits, which means that "a line of thirty
cubits" would not "encircle it completely," but would
necessarily fall 1.4 cubits short. (A cubit, by the way, is
the length of a forearm; the distance from the fingertip
to the elbow, or *cubitus* in Latin – about a foot-and-a-
half.) Again, not a problem if you think that this was
written by people who had a crude grasp of geometry
and knew that π was about 3, and that cubits weren't
particularly precise measures. But if you think it is
the Precise Inspired Literal Truth, then the passage is
problematic, for it directly implies that either the diameter is given incorrectly, or the circumference is given
incorrectly. Did God not know the value of π? Or was
geometry miraculously different for King Solomon? The
rationalist suspects that God, if He exists, knows the
value of pi; but the author of the passage didn't.

Or consider this: According to Genesis 3:14, God
curses the serpent in the Garden of Eden and condemns
it to locomote in the familiar fashion that we associate
with snakes: "upon your belly you shall go, and dust
you shall eat all the days of your life." Snakes indeed go
upon their bellies, but of course they do not in fact eat

dust. You can look it up in a herpetology textbook. So part of the curse apparently works, the belly part – and part of the curse doesn't work, the dust consumption part. This is not a problem for a rationalist, who sees "dust you shall eat" as a figure of speech indicating a diminished status associated with crawling on its belly. But it poses a bit of a problem for the literalist, who is obliged to explain away the unsuccessful half of the curse, or acknowledge that the passage is not actually to be taken at face value – which is, once again, precisely what the literalist denies. After all, if "eating dust" is a figure of speech, then you might as well just take the "six days of creation" similarly.[3]

Or this: We all know that Judas betrayed Jesus, but how did Judas expire after that act? Matthew 27:5 tells us "Throwing down the pieces of silver in the temple, he departed; and he went and hanged himself." That does sound rather straightforward. At least until you get to Acts 1:18, which tells us that Judas "acquired a field with the reward of his wickedness; and falling headlong, he burst open in the middle and all his bowels gushed out." In this version he sort of explodes, rather than committing suicide by hanging. Is there any way for the literalist to reconcile these two stories, aside from arguing that he hanged himself, then exploded? To the rationalist, it's just a case of two different literary communities and traditions coming up with different answers to the same question.

Finally, consider this: Matthew's gospel is concerned to give Jesus an appropriately Messianic ancestry, and gives his genealogy through King David, and all the way back to the patriarch Abraham. Matthew 1:16 brings it to Jesus, with "Jacob, the father of Joseph, the husband of Mary, of whom Jesus was born." So Jesus had a father named Joseph, who had a father named Jacob (just like the patriarch Joseph in Genesis had). Luke has a similar, but slightly different, goal: to trace Jesus all the way back to Adam, of whom Jesus could be regarded as a second version, and he also goes through King David, but comes out a different way, through David's son Nathan, rather than through David's son Solomon. Then all the names are different – and Luke even lists fifteen more of them than Matthew, which adds up to quite a few centuries! – until the two gospels re-converge on Joseph. Thus, Luke 3:23, "Jesus . . . was the son (as was thought) of Joseph, son of Heli." Taken at face value, it would seem as though, at the very least, Joseph had two fathers. Even the early Church father Eusebius struggled with this. To the rationalist there is a simple solution: the gospels were written in two different communities, drawing on two different traditions, each reconstructing the genealogy of Jesus, which nobody actually knew – except that there were two constraints: his father was called Joseph, and he had to be a descendant of King David in order to be the true Messiah. The two gospel authors each took different routes to achieve that narrative goal; one through a paternal grandfather named Heli and the other through a paternal grandfather named Jacob. (Being patrilineal, they didn't much care about Mary's ancestors.) It really is not that big an issue. To the literalist, however, the two paternal grandfathers of Jesus would presumably have to be regarded as yet another inexplicable miracle.

The fact that we are talking about miracles, however, shows that, once again, this is not a scientific discourse at all. It is a theological discourse about how to understand the Bible. Whether you choose to believe a scientific story about where we came from, or some other story, is really beside the point. But if your story of choice is biblical, then you are taking a theological position; and regardless of what you think of the science, presumably you would like to be informed by the best modern scholarship about the Bible. After all, it is one thing to reject modern science in favor of theology, but quite another to reject modern science and modern theology in favor of a perverse and contrarian theology.

In Genesis 3:6–7, Eve gives Adam the fruit, and he eats. The text continues: "Then the eyes of both were opened, and they knew that they were naked." Even St. Augustine, around the year 400, recognized that however literally you want to take this story, it would not sustain a reading in which Adam and Eve had been walking around all this time with their eyes closed, and suddenly now opened them. Even if that's what it says. In fact, *especially* if that's what it says. Clearly the business about the eyes is a metaphor describing their inability to see morally before eating the fruit, and subsequently being able to see the immorality of public nakedness. It would be idiotic to think that prior to their eyes being opened they had been walking around with their eyes shut, bumping into the snakes and trees in the Garden. So it cannot be a matter of whether to take the Bible literally, but of how much literalism to apply, and where to apply it.

Cultural Evolution and the Bible

Take, for example, what is arguably our greatest evolutionary achievement: the ability to make and use fire.

Chimps don't have this ability, for they lack sufficient thumbs and brains for the job. But of course, we are not talking here about an evolved biological feature, but rather a cultural feature that coevolved with our biology.

If you believe in that sort of thing. And you might as well, because it isn't as if the Bible actually has anything to say about the discovery of fire. At least the ancient Greeks thought that fire was important enough to make up a story about it, crediting Prometheus with having gifted it to people, against Zeus's wishes. The Akkadians got fire from gods known as Apkallu. The Bible says merely that Cain and Abel offered sacrifices, and we can infer that those sacrifices would have involved burning. And that would necessarily imply that Cain and Abel knew how to make fire (perhaps having eaten the fruit from the Tree of the Knowledge of How to Make Fire), or else they hung around waiting for lightning to strike. But in the modern age, we can create an alternative story that acknowledges fire as a biocultural development, the product of ages of learning about the environment and what the things in it can do, and one that permits humans eventually to transform the environment itself with this knowledge. Rather than seeing it as a gift, or as something that was there from the beginning, we can instead see the production and control of fire as a gradual and critical process in becoming human.[4] We can base this on ethnographic and archaeological data, which show how fire creates the spatial, social, temporal, and ecological worlds that human hunter-gatherers have inhabited. It creates domestic areas, it affords protection from predators, it detoxifies local foods, and it reduces the reliance on sunlight for illumination and on body hair for warmth – all of which are quite different from the comparable fire-less chimpanzee worlds.

Another great cultural achievement was metallurgy, again independently invented by different peoples, and we have a pretty good idea how. People who use stone tools, such as spear-points, will often harden them in fire. When you heat rocks, eventually you get some of them leaving some remains or slag. These would be metals with low melting points, like copper and tin. The learning curve of metallurgy, then, depends on three variables: what kinds of rocks you heat, how hot you can get your fire, and your level of motivation for learning from scratch about rocks and what you can do with them (after you've learned about fire, of course). For the biblical literalist, the task of reconstructing the origins and development of metallurgy is rather easier. Cain has a descendant named Tubal-cain, "who made all kinds of bronze and iron tools" (Genesis 4:22). Apparently he figured it all out for himself, although this account leaves unanswered just how he communicated his discoveries to the people, for example, of South America.

By focusing so much on Genesis as a biology narrative, creationists and evolutionists alike have managed to deflect attention away from the obvious deficiencies in Genesis as an anthropological narrative. As we have already noted, Adam and Eve are placed in a garden and are expected to cultivate it; there is no conception of a pre-horticultural existence. Their sons are a farmer and a herder – Cain and Abel – whose tensions could be understood allegorically, as representing the long-standing economic competition between farmers and herders. After all, they use land differently – to grow stuff and to allow animals to graze, sometimes on the stuff being grown. In the musical *Oklahoma!* – about rural American life in the nineteenth century – Rodgers and Hammerstein wrote a song about how "the farmer and the cowman should be friends." Clearly the tension between herders and farmers is intense and ancient. Cain was not the only farmer who ever murdered a herder, the story seems to be telling us, merely the first.

But what of the actual history of animal and plant domestication? After all, regardless of what people think of the origins of humans and chimpanzees, it's pretty clear that God did not create corn.[5] Humans created corn – in particular, Mesoamerican humans, and over a period of several thousand years. How do we know this? For two reasons: first, corn can't breed efficiently without human intervention, so it had to be a human cultivar from its beginning; and second, we have the learning curve, by which a grass called teosinte, which is still interfertile with corn, gradually became corn, thousands of miles from the Garden of Eden. We also have mapped the major genetic changes that differentiate teosinte from corn.

God did not make cabbage, kale, brussels sprouts, broccoli, or cauliflower. People did; in fact, all from the same wild mustard plant. And if you insist on arguing about where that wild mustard plant came from, then you must like arguing a lot more than I do, and should probably take a Xanax.

Archaeologists can distinguish domestic crops and animals from their wild relatives because they are subtly, but distinctly, physically different from one another. People, not God, directly altered and effectively made their own plants and animals where they lived. And of course they tinkered with different animals and plants in different places. There were no sheep in Peru, and there were no potatoes in Mesopotamia – I suppose a biblical literalist would have to blame Noah for not dropping them off. So local peoples worked with the animals and plants around them, and over the long term transformed them in often quite striking ways.

But this is the heart of Darwin's brilliant nineteenth-century analogy. Where the natural theologians likened species to machinery – like a watch, whose existence depends on a watchmaker – Darwin instead likened species to domesticated subspecies or varieties. After all, if God didn't create English bulldogs or carrier pigeons or Hereford cattle or watermelons or Honeycrisp apples – because we know that somehow people did – then to understand the origins of those animals and plants, we focus on the "somehow," which renders God largely superfluous to the discussion. People, not God, somehow made breeds, varieties, and subspecies by altering their patterns of breeding over the long term. *And species are more like subspecies than like watches.*

That, however, is neither a scientific proposition nor a theological proposition, yet it has implications for both science and theology. Is a species more comprehensible as a living watch, or as a very large strain (variety, breed, subspecies)? Is there a reliable basis for deciding which analogy is more valuable?

In short, no, there isn't. This is a classic example of what the philosopher of science Thomas Kuhn called a clash of paradigms. These are two radically different ways of thinking about the natural world – is a species more like machinery or a population of organisms? – for which there is no empirical resolution. There is just a deep feeling that a species really is more like a breed than like a timepiece, and if you can't see that, it must be because you're a moron. So let's just agree that, whether or not you are a biblical literalist, a species is more like a breed than a watch. This isn't even an argument about the Bible, because watches aren't mentioned in the Bible. This is just rationality applied to the problem of making sense of what a species is. Are different species more like Rolexes vis-à-vis Timexes, or sheepdogs vis-à-vis dachshunds?

Time's up.

5
Myths of Science and Religion

After World War I, it was difficult to argue that history represented progress, as many optimistic scholars since the eighteenth century had maintained. Today we see history in terms of trade-offs: some things improve, other things get worse; technology adds amazing things to our lives, but creates havoc at a far larger scale. Think of the benefits of nuclear energy and the perils that accompany it; the benefits of social media, and the trolls and Russian bots that inhabit it; your knowledge of how to read and drive, and your lack of knowledge of what is edible in your backyard or how to make a fire.

Technology is improving, and for a good pseudo-Darwinian reason: Without improvement, you are at the mercy of competitors. Technology has become intertwined with science, as the latter has developed over the last few centuries. Our technology improves, and our understanding of nature has improved – at least in the sense of having become more accurate. Other things change in reaction to the technology and the knowledge, but it's not clear that they improve. The physical isolation and loss of privacy that accompanies the age of wi-fi is a price we pay for the marvels it brings us. The privatization of genomics means that it is now more difficult to distinguish sales pitches from discoveries in human genomics, and contributes to the overall deterioration of the presumptive truth content of scientific statements.

In some ways we have become a more just society, but science didn't have anything to do with it; that was down to the humanists. Nor has science done much to reduce inequality. Quite the opposite, in fact. The days of Jonas Salk's polio vaccine philanthropy are long gone; today science is there for investors to make money from. Big Pharma isn't in the business of public health; they are in the business of making money, although the two interests hopefully converge with some degree of regularity.[1] The 2010 bestseller, *The Immortal Life of Henrietta Lacks*, showed the life sciences as a fairly simple tool of neoliberal capitalism, albeit a technologically sophisticated one, as biomedical fortunes were made from the cells of a poor black woman, but somehow the prospect of sharing the profits with her family was deemed to be against the interests of science.

The doctrine of progress is a doctrine of ethnocentrism. By being better than our ancestors who had less technology, we are simultaneously better than our contemporaries who have less technology. That is indeed precisely the meaning this metaphysical proposition came to have when it was popular in the late 1800s: Technology is the engine of civilization, which is manifestly better than non-civilization. What then of the less civilized peoples who still exist? Should they be extirpated? Should they be remade? Should they be sterilized? As long as they were perceived to be inferior in some fundamental way, the less civilized peoples of the world were under constant threat from the tag-team of science and progress. Eventually seeing the cultural others (and the cultural ancestors) as different, rather than as inferior, created a new set of possibilities for them. Today, the late nineteenth-century doctrine of progress is seen as a colonialist rationalization, since the scientific ambition of bringing civilization to the uncivilized peoples was happily aligned with the Euro-American political economy of the age. Acknowledging progress meant acknowledging other peoples' lack of it. And indeed this was the era in which science began to formally query the nature of the savage's primitiveness.

Why Are There Savages?

In 1853, Count Arthur de Gobineau used the Huron as a metaphor for all Native Americans when he wrote sarcastically:

> Thus the Huron's cerebellum contains a germ of spirit quite like that of English and French! Why, then, in the course of centuries, did he discover neither printing nor steam power? I should be entitled to ask this Huron, if he is equal to our compatriots, why the warriors of his tribe did not produce a Caesar or Charlemagne, and by what inexplicable negligence his poets and wizards have never become Homer or Hippocrates?[2]

The possibility that they did not know they were in some kind of a race against Gutenberg and Watt does not seem to have occurred to him. Nor does the likelihood that Gobineau's inability to name their prominent warriors, poets, and wizards might reflect his own ignorance, rather than theirs. Nevertheless, it was indeed a question: Why was the primitive so primitive?

One possibility, favored by (the creationist) Gobineau, was fleshed out in 1868 by (the evolutionist) Ernst Haeckel. Savages were different kinds of beings entirely, different by nature. Gobineau ridiculed the Huron's cerebellum. The biologist Haeckel placed them and himself in different species entirely, *Homo americanus* and *Homo mediterraneus*, the latter having evolved farther than the former. Many scholars disagreed with his taxonomy, but they didn't write books as widely read or as influential as Haeckel's.

But if not a difference of nature, then what other kind of difference distinguished the civilized and uncivilized races from one another? Alfred Russel Wallace, co-discoverer of natural selection, published his own idea a year later, in 1869. The difference between savages and civilized people were not differences of nature; their brains were effectively interchangeable. If the difference lay not in a difference of nature, where then might it lie? Wallace reasoned that natural selection could not create body parts with no utility. Since savages have the same brain, but use far less of it than civilized people do, Wallace deduced that the brain could not have evolved by natural selection. Everything else evolved by natural selection, said Wallace, but the human brain must have had a little bit of divine assistance. Darwin wrote him, "I hope you have not murdered too completely your own & my child."[3]

But if the difference between civilization and savagery lay not in the natural order, nor in the supernatural order, then where did it lie? In 1871, Edward Tylor answered that question a different way, and more persuasively than either the racist evolutionist Haeckel or the spiritual evolutionist Wallace. He did so by formalizing the idea of cultural evolution. Cultural evolution became an intellectual space in between nature and miracle, the domain of human history and human agency. It encompassed neither the organic nor the spiritual, but rather, the "capabilities and habits acquired by man as a member of society." Tylor thus called attention to a third reality, in addition to God's domain (spiritual) and God's creation (natural) – namely people's handiwork, their products, institutions, and activities.

Savages, in other words, were savages because of their culture, not because of their brains or because God chose to deny them His civilizing grace. "Cultural" thus became a second antonym to "natural," the first being "supernatural."

Tylor's introduction of culture as an analytic tool to explain the source of mental differences among peoples seems commonsensical to us today. Yet even now it is sometimes contested in reactionary scientific circles. Given that there are natural, biological (the nineteenth-century term was "racial") differences among peoples, and cultural, historical differences, is it fair to assume that any specific cognitive difference is cultural? Could it not be natural? (Nobody today argues for such differences being miraculous.)

As Tylor recognized, this is a biopolitical landscape. He concluded his 1871 book *Primitive Culture* with the thought that "the science of culture is a reformer's science." The people here do this, and the people there do that, for historical, not biological, reasons. They are not lesser orders of beings; and while not fully civilized, they are civilizable.[4] The biologist Ernst Haeckel and his followers disagreed.

Today our familiarity with transnational immigrants on a large-scale shows the full "assimilation" of anyone anywhere to be an historical regularity and a universal possibility. This reinforces the "psychic unity of man" position, premised on the fundamental interchangeability of all human brains, barring the pathological or the rare mutant genius. The alternative suggestion, that any particular cognitive difference among groups of people is a difference of nature, not culture, is a reactionary biopolitical proposition. It is never "just a hypothesis" that maybe some other peoples' brains evolved to be different. In 1962, it was "just a hypothesis" of the anthropologist Carleton Coon to explain why black children and white children should not be in the same American classrooms. In 1994 it was "just a hypothesis" of the psychologist Richard Herrnstein to explain why social programs targeting black and Latinx families were doomed to fail and should be defunded.[5] In the twenty-first century it is "just a hypothesis" trotted out to scientifically explain not just racial, but sexual difference in cognition as well. As we noted in Chapter 3, a century ago William Jennings Bryan properly rejected the obvious sexism in the idea that sexual selection had made women's brains function more poorly than men's.

The idea that any particular mental difference is one of nature, rather than culture, is known as hereditarianism – and like Haeckel's Darwinism, it is often mistaken for normative evolutionary theory. Obviously politically reactionary scientists will try to lend greater gravity and validity to their ideas by associating them with Darwin.

That is why it is a constant battle for the scientific community to keep Darwin's name unsullied. Jesus predicted that "Many will come in my name . . . and will lead many astray" (Mark 13:6), and he was right about that, and not just about his own name. When racists and sexists occasionally poke their heads out of the sand of science, they invariably do it in the name of Darwin.

In one sense, that is not terribly unexpected. In the wake of the Human Genome Project in the 1990s, advertisers appealed to molecular biology metaphors, claiming that virtues were "in the DNA" of everything from airplanes to home entertainment systems. It's not uncommon today to hear that "Aviation is in our DNA" or "Winning is in our DNA," although of course only viruses and cells have DNA and neither aviation nor winning is composed of them. The claims are evocative of science while simultaneously being scientifically meaningless.

Advertisers believe that fad science is a selling point, so it seems as though the public can't really be as "anti-science" as all that – or else such advertising campaigns would never get off the ground. This also creates a paradox, however, because the consumers for whom DNA is meaningful (because they are "scientifically literate") are also being egregiously misled about the science itself. Cells have DNA; high definition televisions do not, regardless of what the advertisements tell you. Science is literal and concrete; this advertising science is literary and metaphorical.

In fact, the popularity of the modern fad "Paleo diet" would seem to be predicated upon the acceptance of evolution, which ought to be a cause for celebration. But regardless of the popularity of a presumptively

Paleolithic diet, anthropologists will nevertheless gladly explain that (1) Paleolithic diets were local, not uniform; (2) Paleolithic foods are not available today; and (3) they were certainly unlike anything approximating a modern "Paleolithic diet." To eat sensibly is not to eat pseudo-prehistorically. So whatever positive science literacy is evidenced by acknowledging that there actually was an Old Stone Age a million years ago is offset by the fact that the public's presumptive knowledge of it is misleading and false.

This is the great paradox of creationism. Not taking evolution seriously enough is bad, but taking evolution too seriously is also bad. Of course the "Paleo diet" is not particularly threatening to anyone's well-being. On the other hand, racism is indeed threatening, and when underpinned by some perverse evolutionary theory, it is bad for science as well as for society. Eugenics – the ambition of improving society by breeding a better citizenry – is a post-Darwinian concept that has invariably been promoted as a deduction from Darwinism, although it is of course much more than that. Hereditarianism invariably cloaks itself in Darwinism as well. Clearly the problem is more complicated than simply leading the public to evolution: We have to lead the public to a normative and benign evolution. That is to say, we are obliged to differentiate thoughtful evolutionary theory from thoughtless, outmoded, and tangential evolutionary theories of dubious social merit and scholarly value.

There are always conservative ideologues happy to use the naive speculations of scientists about the genetics of human nature and wield them as ostensible evidence for the naturalness and immutability of our familiar social hierarchies. Here, "human nature" is taken to constitute an invisible barrier to social progress, and evolution to be the sorting process that created our social hierarchies. By this logic, patriarchy is natural, and racism is unrelated to racial inequality. Moreover, if you don't agree with it, then you must be an anti-science creationist; this argument is familiar in the works of modern conservative apologists like Charles Murray and Andrew Sullivan.[6]

So too with the doctrine of progress. After all, surely our grandparents were less knowledgeable than we are today. Surely our pre-scientific ancestors were far less advanced than we are. Once upon a time (goes this pseudo-history) there was only belief in magic and superstition; then religion coalesced and people got a bit smarter; and now all that has been superseded by science. This was indeed the view held by anthropological science in the nineteenth century, and it is why we regard Edward Tylor as a pre-modern anthropologist, rather than a modern one. For it is hard to deny that magical or superstitious thinking is still very much with us – certainly a visit to a casino or a sporting event ought to convince you of that. Perhaps it is just human nature to seek a measure of imaginary control over things that you can't really control.

Nor does religion appear to be on the wane. Like science, it adapts and evolves. Like magic and science, we compartmentalize it. Someone who is priestly 24/7 is just as tedious as someone who is scientific 24/7.

Science and Religion

The replacement of religion by science has been an aspiration since the French Enlightenment, and has been invoked quite regularly by demagogic scientists since the mid-1800s. The sociologist August Comte, the biologist Ernst Haeckel, the geneticist Francis Galton, and the modern biologist Richard Dawkins all share a common ambition: to replace religion with science. People simply aren't rational enough, dash it all! If only they would listen and adopt more rational sensibilities! We are the future! Resistance is futile!

But in the age of Brexit and Trump, it is difficult to see how anyone could still seriously adhere to the doctrine of progress. In fact, where we have actually achieved a measure of social progress (for example, with the Civil Rights Movement of the 1960s or the end of Apartheid in the 1990s) it was hard-fought, and hardly automatic. Indeed, to reduce social progress to an automatic outgrowth of the mysterious forces of history is to fail spectacularly to credit the people who actually made it happen. The Equal Rights Amendment for women in America was not ratified in the 1970s, and there has not been much impetus to revive it. Has there been progress in gender equality since the 1970s? Of course, but it has been due to hard work, not the magical forces of history.

The doctrine of progress was rejected a hundred years ago because World War I made it seem inane. The more recent development of the (scientifically driven) capability of civilization to self-destruct, and the narrowness by which it has repeatedly avoided it up to now, appear to confirm the case. Indeed there is a parable of the inability of scientists to think in other than technological ways in the famous first meeting of physicist Robert Oppenheimer, head of the Manhattan Project, with President Harry Truman, after the end of World War II. Oppenheimer approached Truman, expecting commiseration, saying, "Mr. President, I feel as though I have blood on my hands." Truman angrily explained to him that he had merely supervised the bomb's construction, but did not have to make the actual decision to use it to kill people. It had apparently never dawned on Oppenheimer that the decisions he made might not have been the most important ones – the moral and political ones. The decision to obliterate the city of Nagasaki and its civilian inhabitants was very ambiguous as a moral advance, while nevertheless unambiguous as a technological advance; fortunately for Oppenheimer, he only had to worry about the latter. For the important decisions, he was just a second-guesser and a nuisance – in Truman's words, "a crybaby scientist."[7]

The doctrine of progress, though, is also naively accepted by many scientists today. How could that be? We can be charitable: it's about metaphysics, which isn't their bailiwick. And they experience it in their science, which is what they are immersed in. Moreover, it places them at the apex of history: Magic was supplanted by Religion, which is now supplanted by . . . Me.

This is an unfortunate situation, since it reproduces one of the more familiar hallmarks of colonialism: Your ideas are wrong and ours are better, so you had better accept them. Again somewhat paradoxically, there has been a bit of progress toward a post-colonial world, but not with this kind of attitude!

The colonial mindset is an old and familiar one, and certainly not one a scientist ought to be proud to have. There are two qualifiers, though, on either side of the fence. On the one hand, science is indeed improving our knowledge. On the other hand, bound up in the advancement of science have been embarrassments –

grave-robbing, scientific justifications for genocidal practices (Europeans had such big skulls, after all), and other forms of complicity with, and rationalization for, the colonial powers. So when apologist Steven Pinker tries to downplay the notorious Tuskegee syphilis experiments on the grounds that the poor black subjects were not actually infected with the disease by the doctors (they were merely untreated, to see what the untreated disease would do to a human body),[8] he neglects to mention the fact that there were indeed people that the U.S. Public Health Service thought so little of that they actually *did* infect them – men in jail in Guatemala.

All of which is not to bash science, but rather to admonish it against adopting an archaic political worldview. We now recognize colonialism as bad, something we confront and would like to transcend. Modernizing and remaking the natives so we can take advantage of them is not considered admirable behavior any more.

But surely science is right and therefore should replace other worldviews? Certainly sometimes – a good example is the eradication of a prion disease called kuru that afflicted only a small population of New Guineans, and was defeated by imposing the cessation of cannibalistic funerary practices. But we might also turn the question around and ask: What motivates scientists to reflexively combat or degrade other modes of thought?

Religion is, after all, more than merely a set of competing narratives about how the universe works. There is indeed such a rational component to it, producing explanations of how and why things happen, but ones that are inevitably less accurate than scientific explanations, if that is your standard for such narratives. But religion is also performed ritually, experienced emotionally, and partaken socially; and that is actually quite a lot to expect to supplant with just a more accurate story.

If religion is more complicated than it may seem at first glance, so too is science. Science is more than just a set of methods to generate increasingly accurate stories about how the world works. Science is a voice of authority in the modern world (and you probably know how voices of authority reflexively respond to having their voice challenged). Science is also a guild, requiring many years of study to enter; and an occupation, with the attendant insecurities. It also has a grey area around it: Is neoliberal corporate science really science? Is what's good for Monsanto necessarily what's good for science? Is direct-to-consumer genetic ancestry testing science? Is scientific racism science? Were phrenology and eugenics science? Are evolutionary psychology and exobiology science? Is math science?

Armed with the knowledge that (1) technology and scientific knowledge improve and other cultural forms adapt to them without necessarily progressing, and (2) science represents a different mode of thought than religion, we may return to the question posed in the title of this book. Creationists are not holdovers from an older, stupider time. They are reactionary, not primitive. They have adopted an extreme theology in the face of rationalist narratives about who we are and where we came from. Their theology is consequently what defines them, not their approach to science.

Anti-Anti-Science

Indeed, thinking about creationism as part of an "anti-science movement" obscures its relationship to other "anti-science" positions. About 10 percent of Americans oppose vaccines, while about 40 percent oppose evolution – and the extent to which those numbers may overlap is unknown. The few millions of Americans who oppose vaccination generally do not hold those views for religious reasons. The anti-environmentalism of many Americans is likewise driven primarily by corporate greed, not piety. What makes creationism unusual among "anti-science" positions is in fact its theological basis.

The solution to creationism, then, is not better biology; it's better theology. Or to put it another way, unlike other "anti-science" positions, which can (at least in principle) be debated and adjudicated by data and reason, creationism is more fundamentally a disagreement about how to understand the Bible. To the creationist, the Bible is not a compilation of selected texts rendered sacred over hundreds of years of recitation, redaction, translation, and interpretation, but rather a singular corpus of diverse truths, decontextualized and abstracted from history and culture, to be situationally invoked in the modern world. The problem actually lies not in the creationists' confrontation with science, but in their confrontation with the Bible.

Indeed, creationism is one of several social issues that need to be understood as theological disputes in order to be productively addressed. A related issue is known as the "prosperity gospel," an American-born message that God wants you to be pious and the currency of His piety is literally financial. So you give your money to God (or at least to His earthly ambassador) and He will reward you, and you will experience His grace in your bank account. On the face of it, this message would seem to be difficult to reconcile with some of the most famous sayings attributed to Jesus: "It is easier for a camel to go through the eye of a needle than for someone who is rich to enter the kingdom of God" (Matthew 19:24). "You cannot serve both God and money" (Matthew 6:24). "If you wish to be perfect, go, sell your possessions, and give the money to the poor, and you will have treasure in heaven; then come, follow me" (Matthew 19:21).

Nevertheless, the construction of an American alt-gospel, in which Jesus actually likes rich people and wants you to follow him in gold and diamonds, or in the expectation of gold and diamonds, has been an exceedingly popular message for decades. Unsurprisingly, some of its exponents have amassed fabulous wealth in contributions from followers. Some of the most egregious have served jail time for fraud, notably televangelist Jim Bakker; but others – most famously, Tulsa's Oral Roberts, and more recently Houston's Joel Osteen, Atlanta's Creflo Dollar, and Dallas's Kenneth Copeland – have only had to tolerate the passing scorn and transient ridicule of their evangelical Christian peers. This does not seem to have bothered them very much, nor their many followers. Indeed, the "prosperity gospel" is a significant social and religious movement. Nevertheless, its battleground is theological: What kind of religion is this, which reverses its founder's clearest message and shamelessly sends its worshippers into hock for the sake of its garishly avaricious leaders?

Likewise with praying, another theological issue with social implications. Praying is fine, but not as a substitute for acting. In praying for a favorable outcome instead of working to bring that outcome about (such as legislating ownership of assault rifles) the worshipper seems to confuse God with Santa Claus, as theologians have indeed noted.[9] Actions do speak louder than words, especially when the words are addressed to heaven, rather than to other people on earth; and praying-instead-of-acting has obvious political consequences that are not lost on the more cynical advocates of praying.

Similarly, the extraordinary support (70–80%) among white evangelical Christians for the most mendacious, disreputable, corrupt, and morally bankrupt American president in memory stands as a stark theological issue. The disparity between white and non-white evangelical Christians' support for the Trump administration suggests the existence of a theology among the white evangelicals that is more accommodating to racist rhetoric and policy than we have been accustomed to expect in the modern age. If morality is a pillar of the Christian religion, or ostensibly of any religion, then what do we make of a morality that embraces and exalts a serial philanderer, bald-faced liar, and nepotist – and yet which only a few years earlier had ostentatiously invoked "personal character" as a critical political trait? Why should the protections for automatic weapons and fetuses outweigh the protections for civil rights, "family values" and personal integrity that once predominated in American political discourse? Once again, a finger-wag at the obvious hypocrisy seems hardly adequate; this is again a theological issue. Where is the moral Christian basis for admiring the basest of American scoundrels, or for racism?[10]

We don't know the precise overlap between Trumpists, prosperity-gospel advocates, white supremacists, and creationists. Nevertheless, seeing creationists as simply "anti-science" hardly gets to the nub of the matter. Science is marginal here, even to the extent that environmentalism and climate-change denial also cross-cut the politics. The real issue is their engagement with morality and with, for lack of a better term, their deity.

I am not arguing here on behalf of no deity at all, but rather, against this particular one – the deity who wants you to do the opposite of what He tells you; who dictates morality but enjoins you to embrace immorality; who preaches love but mandates intolerance. There are, of course, many conceptions of the deity out there, with many attributes ascribed. But this one is seriously creepy, and creationism seems to be only one of His quirks. Again, the battleground here is theological – concerning the nature of God and His moral precepts – not scientific. Science is only a distraction.

Science and the Moral Realm

Whether or not religiously founded, today we recognize morality as a product of cultural history – the object of a kind of evolution. Although many of our modern political, social, and moral ideas were originally founded on the Bible, today we believe very differently than they did in ancient biblical times. They believed in polygamy, slavery, spirit possession, witches, and monarchy; we believe in monogamy, freedom, germs, mental illness, and democracy – clearly a different and unfamiliar set of ideas and values with which to confront a Jew or a Christian of many centuries ago. Thus, morality is historically produced and situated, and while compatible with a divine origin, it nevertheless doesn't require a heavenly source in order to be normative and coercive.

Every adult everywhere is enjoined to do what is right and good and expected, and to eschew the opposite, or else risk the consequences, which may be earthly or cosmic. Certainly rule-governed behavior is an essential attribute of all human societies, and whether the primordial rule was "Don't have sex with your sister" or something else, moral codes of conduct are ancient, ubiquitous, diverse, and fundamentally human.

Like any other human activity, science must be good, or else face the consequences – the traditional imagery is generally torches and pitchforks, but the modern reality is loss of funding. Science can no more be amoral than Adam and Eve could after eating the fruit. That story seems to tells us that amorality – a human existence without right and wrong, without good and evil – is simply no longer an option for an adult. Of course, we can debate what falls into either category, since the only specific thing the Bible tells us that Adam and Eve internalized from their newfound knowledge of good and evil was not to be naked in public. And that still sounds reasonable.

A final element of the Genesis narrative, often overlooked, is the role that morality plays in the cosmic order. God is, in fact, quite explicit about it: "See, the man has become like one of us, knowing good and evil ..." (Genesis 3:22). Apparently to the ancients, morality itself is what separates God from His creatures; but having eaten the forbidden fruit, people are now like God, specifically in knowing good from evil. And it is hardly a stretch to see the adherence to abstract, symbolic, and often arbitrary rules as a primary zoological difference between people and even our closest relatives, the apes.

But that very mental property also allows us to turn our gaze reciprocally upon God, and to query His own morality, as many biblical commentators have long noted, and indeed attempted. (Although when Job did so, somewhat later in the Bible, God asked him rhetorically, "Where were you when I laid the foundation of the earth?" In other words, STFU.)

And yet, we know that God is fundamentally a moral being. After all, He said so, back in the Garden of Eden. The God of the Bible is not like Loki, for example, a trickster who is forever confounding his brother and rival, Thor. Indeed, the trickster God is a familiar figure in many pantheons of diverse peoples. But that is not the Christian God, of either testament.[11]

This is an important theological point, too. Back in 1857, just two years before Darwin published *The Origin of Species*, a pious naturalist named Philip Henry Gosse published a book called *Omphalos, an Attempt to Untie the Geological Knot*. The knotted problem in geology was that the world seemed to be very very old, in spite of the Bible seeming to say that it was created only a few thousand years ago. Omphalos is Greek for bellybutton, and Gosse reasoned that Adam had one, even though he had had no mother or umbilical cord. But since Adam was a normal person in every respect, God must have made him with a bellybutton that he had never actually used, a feat of anatomical legerdemain. Likewise, argued Gosse, the earth only looks old because God made it to look old; however, it is really young, as the Bible indicates. So God is actually fooling us with geology, as he fooled us with Adam's anatomy.

Mid-nineteenth-century Christian theologians were unimpressed. If you think God made the universe to look old, as a cosmic illusion, for some bizarre and unfathomable reason, then you are faced with the nature of the deity you have created by such thinking. A contemporary theologian called this god "Deus quidam deceptor" – a sometime liar-god. While the "tricksters" appear in the mythologies of many societies, such a god is not one that nineteenth-century Christian theologians deemed worthy of their admiration, much less veneration; and it certainly did not represent the wise and loving Creator they saw in nature.

Today, when we sequence DNA, we find an interesting pattern that calls for an explanation, just as the rocks did 200 years ago. We compare the sequence of DNA bases – the precise series of A, G, C, and T – across species, and of course the DNAs of species that look anatomically similar are also genetically similar. But within that trans-species genetic similarity, a very non-random pattern exists: across species, the DNA *within* the genes, the functional bits of DNA, are always a little bit more similar to each other than the rest of the genome, the DNA *between* the genes, is. Why should that be? Why should the DNA of genes possess fewer mutational differences than the DNA not of genes?

Actually, the answer is well known. What differentiates genic DNA from nongenic DNA is that the former has a physiological function – classically, to make a protein, but now conceptualized a bit more broadly to have any sort of recognizable function. Any mutational change to a gene may affect its function, and would most likely corrupt it in some fashion, since the gene was already working fine when the mutation occurred. (That is why mutations give you cancer, not X-ray vision!) Such a genetic change would thus have an adverse effect on the organism, and would likely be "weeded out" by natural selection. A mutation in an inter-genic region, however, would not have such an adverse effect, because that stretch of DNA doesn't do much; consequently, it would not be weeded out, since it would not be expressed in the first place. It would simply stay in the DNA, with other such harmless mutations. In other words, we have a good evolutionary explanation for the non-random pattern we see when we compare DNA sequences across species, with more mutational differences in non-genes than in genes. It is a consequence of a long history of small changes in DNA regions that are pruned because of their adverse effects on genetic function, or not pruned because they don't have much of an effect.

But suppose you don't believe in mutations or selection? What sense could you make of that pattern of difference when you compared the genomes of, say, a human and an orangutan, and you always found greater difference between genes than within genes? Without evolution, that pattern would have had to be inscribed in the DNA at the time of Creation. It would be a pattern that strongly suggests selective pruning following random DNA alterations, but of course it would not actually be that at all. It would be merely a deception employed by God to make it look as if evolutionary processes had been at work when they really hadn't. And this is precisely the same theological problem faced nearly two centuries ago with the conception of a "trickster" god who would be more at home in a different culture, and who was rejected for that very reason by thoughtful Christian scholars confronting patterns in the natural world. That just isn't our God; it is someone else's.

Creationism is an intellectual problem for theology, not for science. Indeed it aligns science with one side of a theological argument: for a rationalist, scholarly understanding of the Bible, and against an ahistorical, selectively literal understanding. One can, of course, acknowledge the Bible's sacredness, while at the same time historicizing it. But understanding creationism as a theological rather than a scientific issue affords a different perspective than we usually see. It's not that creationists privilege the Bible over science, but that they privilege a specific kind of engagement with the Bible. To argue over whose narrative of human origins is more accurate is to miss the point entirely. At issue is how one constructs origin narratives.

Scientific origin narratives are constructed with the goal of expressing maximum accuracy. That is of course not necessarily the primary goal of other narratives, which may also be affirming the listener's place in a benign universe, establishing membership in a social and moral community, or may simply rhyme.[12] It is also a mistake to understand creationists as being merely anti-science, as if creationism were part of a broader conspiracy. This is misleading for three reasons. First, the position of science is often more ambiguous than it is in relation to anthropogenic climate change and evolution – consider the ostensible position of science on genetically modified foods, nuclear energy, the innateness of intelligence, or abortion. And second, some science should indeed be rejected – for example, racist science, unethical science, and science with financial conflicts of interest. Accepting everything a scientist says would be entirely uncritical, the very opposite of a good science education. And third, we don't know that there is any relationship at all between, say, the creationists and the anti-vaccinators.

Creationists are thus neither basically pre-scientific nor basically anti-scientific; they are simply selectively biblical. The most salient feature of creationism is its radical theology. Whether our evolutionary narrative of our origins is more accurate than theirs is the wrong question. Science is most fundamentally a system of tools for producing narratives of maximum accuracy; so of course the narratives it produces will be more accurate than the alternatives. The proper question, rather, concerns how we produce these narratives. What assumptions form their basis, and how do we think about history, both our own and that of our sacred objects and beliefs? These are questions that lie beyond the realm of science – in the humanistic domains of philosophy, history, and religious studies. As such they are complementary to the science, not antagonistic. The science is incomplete without them.

6

Sacred Ancestry

A favorite creationist "gotcha" question is: If we evolved from apes, how come there are still apes? That's actually an easy one; after all, we evolved from creationists and yet there are still creationists. One has to stop thinking vertically to make sense of diversity; there simply isn't, and never has been, a Great Escalator of Being. But just as the apes we evolved from millions of years ago were different from the apes of today, so too the creationists we evolved from in the nineteenth century were different from the creationists of today. Today's creationists are more radical and less in tune with the study and understanding of Scripture than at any time previously – either pre- or post-Darwin.

There are many great mysteries out there – such as the nature of consciousness, why there is anything instead of nothing, where the universe came from, why we should be good, how to be happy, the ending of *2001: A Space Odyssey* – but among them is not "Where did we come from?" We came from apes, over the course of the last few million years.[1] We know this by virtue of the standards of knowledge production, which establish how we know anything about anything.

Some of those questions are weird or trivial. I neither know, nor care, where the universe came from, and I hope that the answer to that question will arrive with the answer to achieving the justice, mercy, and equality that we actually need. But if we did not evolve from apes over the last few millions of years, then nothing is actually knowable. The moon landing was sheer luck, antibiotics are equivalent to prayers, and Napoleon is just as likely to have won the Battle of Waterloo as to have lost it.

Let's assume, instead, that there is such a thing as knowledge. The fossil record and our pattern of similarities to living apes strongly indicate that our ancestors were apes. Nevertheless, within the constraints of knowing that we evolved from apes, that our bodies are modified ape bodies, there are great mysteries that remain. Was the emergence of our species foreordained, as part of a plan? If not foreordained, then was it mostly a matter of superior genes or of ridiculously good luck? And if both, then how were they distributed? Does the same thing that happens to humans after they die also happen to chimps? What motivated our early ancestors to stand/walk/run on two legs rather than on four? Were early people perpetually on the verge of starvation? If so, why did they make particular edible foods taboo? And if not, then was natural selection such a serious factor in their evolution? How did they figure out what was edible, or how to prepare food so it became edible? Or were early people perpetually poisoning themselves, in which case, wouldn't that have been singularly maladaptive? Where did early people live when there weren't caves available? Why does the hair on our heads grow continuously, but not the hair on our bodies? Why do we have more hair in our stinkiest places? What can our brains do at a 1500 cc average that they couldn't do at a 1000 cc average? And why can someone today with a 1200 cc brain think exactly the same thoughts as someone with an 1800 cc brain?

The Power of Ancestry

In pop culture in 2019 (spoiler alert!) the *Game of Thrones* Dragon Queen Daenerys Targaryen surrendered to her family's mad blood-lust, while the *Star Wars* Jedi Rey Palpatine rejected hers. The fact that these were culminations of important story arcs shows just how culturally significant the idea of ancestry is. Indeed, the blood-lust of the ancestors even arises in scientific narratives. The anatomist Raymond Dart, who first identified the fossils of *Australopithecus* in the 1920s, came to see them in macabre terms, as "carnivorous creatures, that seized living quarries by violence, battered them to death, tore apart their broken bodies, dismembered them limb from limb, slaking their ravenous thirst with the hot blood of victims and greedily devouring livid writhing flesh."[2] He imagined the wars up to and including the mid-twentieth century horrors to have been a passive inheritance from the Pleistocene.

And perhaps they were. But that is a highly value-laden narrative, which removes any responsibility for war from the decision-makers and participants, and absolves them of any moral judgment. And given that all we possess of those creatures whom Dart was talking about is their anatomy and a bit of their chronology and ecology, it seems gratuitous at best to describe them in such a fashion. Modern evolutionary narratives about *Australopithecus* generally don't depict them as ferocious and aggressive, and modern evolutionary narratives of war generally don't envision it as primordial and natural, and certainly not as horrid as Raymond Dart did.

Aside from the anatomy, chronology, and ecology, then, what we say about our ancestor *Australopithecus* is as mythological as a story about snakes and apples[3]– except that our ancestral stories are constrained by our knowledge of what apples, snakes, and apes are capable of doing. Fruit can't make you self-aware, snakes can't talk, and apes can't fly, so any story that has moral fruit, talking snakes, or flying apes simply isn't a good scientific one. Scientific stories are constrained by knowledge. Creationist stories, on the other hand, simply become knowledge by defying or rejecting the scientific virtue of empiricism, which is what makes stories about human evolution "self-correcting."

The empirical constraint of science is connected to the modern constraint of materiality in naturalistic scientific explanation. That is to say, not only are our ideas rooted in data or evidence, but in a particular kind of evidence that can be shared, independently observed, and can convince an open-minded skeptic. Indeed, the constraints on modern explanations transcend natural science and govern the humanities as well. Where centuries ago history could be written as the will of the gods, we now explain history in terms of groups of people doing things. Their motivations to act are more important than the will of the gods. What the gods may have had in mind, or how they may have assisted, are no longer within the purview of the historian. It's not that those thoughts are now unthinkable, just that they simply don't get thought within the context of modern scholarship about history.

And yet, thinking about our ancestors is not quite so straightforward. Ancestors are where the modern symbolic boundary between the material and immaterial realms – between nature and supernature, matter and spirit, process and miracle – breaks down.

In anthropology, ancestor worship lies at the nexus of the study of kinship and the study of religion. In its most literal sense, ancestor worship is known ethnographically in many parts of the world. In its broader sense, however, treating the ancestors as special or sacred without necessarily worshipping them is universal. Human newborns are special, because of the nature of human evolution that makes human birthing so different from ape birthing (see Chapter 2). Human birth is social and ritualized; death can be considered its symmetrical, social, and ritualized opposite. Human corpses are sealed, burnt, buried, mummified, eaten, or treated any number of different ways.[4] They are regarded as special or sacred – and have been so for possibly hundreds of thousands of years. The dead are treated with respect, even though it doesn't matter to them.

Or does it?

Even though great-grandma is decomposed and skeletonized, we still don't want anyone exhuming her, defacing her grave, or treating what's left of her as an "it" rather than as a "she." Chemically and biologically, it may not matter, but this isn't the world of chemistry and biology; it's the world of symbolic human thought.

Sacred Ancestry

In fact, not only do we think of great-grandma's remains symbolically, but we specifically reject the scientific understanding of her body as simply dead meat and osteological specimens. No, great-grandma is a "person"; her spirit may not live on, because we may not believe in spirits, but we insist on treating her remains respectfully. Why? Because she was once alive; she will always be a person and an ancestor. And personhood and ancestry transcend science. That indeed was the legal recognition behind the U.S. law known as NAGPRA, the Native American Graves Protection and Repatriation Act of 1990. The scientific status of Native American remains is outweighed by their humanistic, symbolic status.

And that, I think, is the key to understanding the persistence of creationism. The ancestors are sacred. They may be ghosts, or corpses, or fossils, or a naked couple in a garden, but the idea that you are part of a lineage is a powerful and universal one, full of symbolic energy. Consequently names get changed to conceal ethnicities, heirs get disinherited, grandparents turn over in their graves, and having one non-white great-grandparent made you non-white in most U.S. states a century ago. That all of this might be relatively impervious to science ought not to be so surprising or incomprehensible.

We see the small-scale negotiation between scientific and non-scientific ideas of ancestry in the modern fad of genetic ancestry testing. A company like 23andMe analyzes your DNA, compares it to their database, and tells you that you are 13 percent Irish, 48 percent Dutch, and 39 percent Tibetan. If you like your results, you can believe them; and if you don't like your results, you are free to disbelieve them. The company tells you up front that they won't stand up legally. White supremacists who don't come out as European as they'd like can cite plenty of methodological weaknesses in the analysis. The internet is full of stories of anomalies, like identical twins whose ancestries are similar, but divergent. The internet is also full of people who take their results so seriously that they start wearing kilts in honor of their newfound Scots ancestry, or develop a palate for kielbasa on account of their new Polish heritage. It's science all right, but a new kind of science devoted to selling stories about ancestry, which are generally kind of true, and usually close enough that it doesn't matter if they're wrong or by how much. What is new and unusual is that people will pay good money for an ostensibly scientific narrative about their individual ancestry, carrying some degree of cultural authority, which is why the industry is booming.

Genetic ancestry testing isn't like creationism. It is performed by scientists as a moneymaking neoliberal capitalist enterprise; the companies themselves encourage you to believe their scientific results but won't defend them in a legal context; and the results might be correct even though there are plenty of scientists telling you not to place too much stock in them. Consequently, you are indeed free to reject the answer you receive, unlike evolution or vaccinations or climate change. But the tests do demonstrate the importance of ancestry as a crucial aspect of constructing your identity, as a means of situating yourself in symbolic social space.

Or consider the famous "Peking Man" fossils, excavated in the 1920s and 1930s from a cave site now known as Zhoukoudian. Although faithful casts of the fossils were made, the original specimens were lost by the Americans in 1941, as they were being transported to ostensible safety from the Japanese. But why were the Japanese so interested in Chinese Pleistocene fossils at all? The Nazis had little or no interest in the Neanderthal fossils housed in France and Belgium, although they certainly recognized the value of archaeology in national myth-making. Ancient hominin fossils from Palestine and Greece were known, but not considered to be in any special danger from the Axis powers. Why did the Chinese fossils need to be protected from the Japanese?

The reason is that since 1937, the Japanese had been making no, er, bones about the fact that they intended to find, confiscate, control, and perhaps even destroy the Peking Man fossils. They had already visited the site and killed people there. The issue was only tangentially about science. It was really about unmooring the Chinese from their ancestors and taking possession of the ancestors, which was far more important symbolically than the mere science – and particularly in Asian cultures that widely hold ancestors in reverence.

There is a direct analogy between the loss of the Peking Man fossils and the replacement of the Genesis story with Darwinism. The ancestors are gone and cannot be recovered, much as we may wish they could be. The ancestors live on in symbolic form as transformed metaphors, but no longer as actual ancestors. They have been disconnected and untethered, now drifting about in metaphorical space. The ancestors aren't coming back, except in an immaterial, and possibly angry and vengeful, state.

In this sense, then, creationism answers the question: Who ya gonna call?

Creationism is an attempt to reclaim the ancestors, to reconnect to a particular sacred lineage. And like other ties to remote ancestors, it is more symbolic than genetic. After all, even Peking Man may not have been literally your ancestor. When we say Peking Man was an ancestor, we mean that something or someone anatomically very much like those fossils was your lineal ancestor; quite possibly, some members of *Homo erectus*, broadly defined. But we don't even really know if Peking Man was the literal genetic ancestor of anyone in Beijing today. Lineal ancestry gets very confusing when you go back just a few thousand years. Since every ancestor had two parents in the previous generation, if you go back, say, fifty generations, you must have had 2^{50} ancestors in that generation, or more than a quadrillion, which can't possibly be right.[5] There is always a lot of symbolic shorthand going on in any conversation about ancestry.

Conclusions

Creationism, especially in its most recent version, Intelligent Design, is an antagonistic ideology, standing for nothing, but against the idea that we are the genealogical descendants of other species over the course of terrestrial history. Of course we seek meaning in our lives. Humans are built for meaning-making; we seek it out, and when we don't find it, we make it up. Making meaning is a creative task for mythmakers, storytellers, theologians, ethicists, and philosophers. In fact, since scientists study nature, not meaning, and nature isn't inherently meaningful, it's a bit paradoxical that the people actually least qualified to make those meanings are the scientists themselves. It's a humanities job.

That's why we grimace today at the writings of

geneticists in the 1920s on sterilizing the poor, and at psychologists writing on the intelligence of the poor. The meaning of poverty, inequality, and prejudice is not something that the study of nature can illuminate, for they are not natural facts, like the meaning of frog metamorphosis. They are political facts. Also unlike the meaning of frog metamorphosis are the meanings of life, of death, of evil, of dreams, of love, of honor, of faith, and of the Bible. Scientists are not particularly good at any of those. Science can explain death biophysically, but whether this actually explains death – that is, whether a biophysical explanation is the kind of answer we seek – is contestable. But we don't deny that death is real, a phenomenon necessitating an explanation. To explain evil, we don't begin by rejecting its existence. We first recognize it.

Likewise, to explain the meaning of our ancestry, we don't begin by denying our ancestry. We grapple with the meaning of our physical descent from apes by acknowledging it as something meaningful. If the meaning we seek is theological, then we grapple with it theologically. But that is exactly what creationists don't do; their theology rejects the question rather than trying make sense of it. It imagines science to be a competitor, rather than a complement, to the humanistic world of semiotics and hermeneutics, of meaning and interpretation. Philosopher/theologian David Bentley Hart ridicules this position with an analogy:

> Christians believe God is the creator of every person; but presumably none of them would be so foolish as to imagine that this means each person is not also the product of a spermatozoon and ovum; ... God's act of creation is

understood as the whole event of nature and existence, not as a distinct causal agency that in some way rivals the natural process of conception.[6]

You are free to consider dear old Dad as your father in one sense, and God as your father in another sense. Only the first sense is scientific, but both may be important. Paleontologist Stephen Jay Gould attempted to demarcate the respective roles of science and religion as "non-overlapping magisteria," essentially reserving science for facts and religion for values, and seemingly embracing all humanistic endeavors within "religion." But actually only science is bracketed, as it was constructed to grapple specifically with law and process in nature. It established a domain for itself within overall (European) knowledge beginning in the sixteenth century. Religion was already there; indeed theology was known as "the queen of the sciences." It was the "new philosophy," as early modern science came to be known, that had to carve out a niche for itself, produce accurate descriptions of the splendors of Creation, and claim authority within its boundaries. It was never intended to keep religion out, just to keep the empirical study of nature in.

Unfortunately, it is the scientists themselves who persistently trade on their authority in the natural realm to try and pronounce authoritatively on the non-natural realm. Historian Jacques Barzun noticed the problem decades ago:

> Where and what am I, whither bound and for what ends? These questions that man keeps asking, all agree that science cannot answer. But the confirming cliché – Science tells How not Why – is falsified in reality by the appearance

of answer-giving which science has been guilty of for over a century. And when it has not so transgressed it has issued prohibitions against answers given by others.[7]

To think perceptively about values and meanings should not be an alternative to thinking about science; it should be a part of thinking about science. There is a popular contemporary meme that goes, "Science can tell you how to clone a Tyrannosaurus rex; Humanities can tell you why this might be a bad idea" – as if to try and explain to scientists why knowing about right and wrong is valuable. Of course, that knowledge was considered so valuable by the ancients that they made it the centerpiece of their story about where they came from. But knowing right from wrong is not a centerpiece of science; indeed, now scientists apparently need to be convinced it is useful knowledge at all. The problem lies in regarding science and humanities (or in this instance, molecular genetics and bioethics) as different things that occasionally come into contact, when they are actually occupying the same intellectual space, but in different ways – like your two fathers. Likewise, modern humanistic biblical scholarship and modern scientific biological scholarship inhabit a common intellectual space, in different ways, as the foil of modern biblical literalist creationism. Creationists reject the work of modern biblical scholars and of Christian theologians as a prerequisite for challenging the biologists.

Or, to put it another way, the fundamentalists' rejection of the humanities scholarship is even more fundamental than their rejection of scientific scholarship.

Of the early Church fathers, the one who was probably most familiar with the biblical text was Origen of

Alexandria, who lived within two centuries of Jesus, and whose major work was a side-by-side comparison of six versions of the Old Testament in Hebrew and Greek, known as the Hexapla. In addition to translating and comparing them, Origen also wrote many treatises about the Scriptures. Like other biblical scholars, he recognized that the meanings of biblical passages must often be teased from their words, and he had no use at all for biblical literalism, even in the year 220. In a famous passage, he asked rhetorically:

> For who that has any sense would suppose that the first, second, third day, with evenings and mornings, existed without a sun, moon, and stars – the first day without even a sky? And who is so foolish as to suppose that God, like a farmer, planted trees in paradise, eastward in Eden, and planted a tree of life of visible and palpable wood, so that anyone biting into the fruit with bodily teeth obtained life? And eating again from another tree, should come to the knowledge of good and evil? When it says that God walked in the afternoon in paradise and that Adam hid under a tree, I think no one can doubt that this is related figuratively in Scripture, and indicates some mystical meaning.[8]

Taking Genesis as a simple historical chronicle wasn't a scientific absurdity in the third century; it was a theological absurdity within Christianity. Now, many hundreds of years later, it re-emerges as a scientific absurdity as well.

There is a joke that goes: A child asks Mom about their ancestry. Mom explains that their ancestors were wonderful and admirable – hardy pioneer stock, staunch abolitionists, a tradition of doctors and teachers extending back beyond the Renaissance. Then the child asks the same question of Dad, who somberly explains that their ancestors were hairy apes and walking fish and ultimately just blue-green algae. Confused, the child goes back to Mom to reconcile the two accounts, and Mom explains, "Well, I told you about my side of the family, and Dad told you about his."

Notes

Preface

1 The word "literalist" can be problematic. Many pious Christians thoughtfully interpret and understand the words of the Bible literally, without imagining it to be a science text. I am using the term more narrowly, to refer specifically to people who reject evolution on the basis of what it says in the Bible. Their theology is actually quite selectively literalist, however, as they tend to ignore or reinterpret other biblical passages (see Chapter 5). Another term often used in this context is "biblical inerrantism."

2 Traditionally, this position is known as theistic evolution, but there is some classificatory confusion. For a notable example, BioLogos.org, founded by Francis Collins, the current head of the National Institutes of Health, proposes that evolution and creation constitute a false dichotomy, and stakes

out a position of "evolutionary creationism". In the present context, however, I am using "creationism" more narrowly, working within the framework of the dichotomy as it is generally recognized, and I classify theistic evolutionists as evolutionists. While I am generally in favor of calling people what they want to be called, for the present purposes I do not regard them as creationists.

3 S. Coakley, 'God and Evolution: A New Solution', *Harvard Divinity School Bulletin*, Spring/Summer 2007, p. 10.

4 "For example, it appears to be inherently impossible for Mars to collide with Venus at some point outside the earth's orbit, as Velikovsky proposes, with the consequence that Venus is knocked into a nearly circular orbit well within the earth's orbit, and Mars remains in a nearly circular orbit outside the earth's orbit. This seems to be the case for the same kind of reason that you cannot pour two quarts of water into a one-quart jar; the world, as far as we can tell, simply isn't built that way." H. Margolis, 'From Washington: Velikovsky Rides Again', *Bulletin of the Atomic Scientists*, 20:39, 1964.

5 E. C. Scott, *Evolution vs. Creationism: An Introduction*. Berkeley: University of California Press, 2005. M. Ruse, *The Creation-Evolution Struggle*. Cambridge, MA: Harvard University Press, 2005. R. T. Pennock, *Tower of Babel: The Evidence against the New Creationism*. Cambridge, MA: MIT Press, 1999.

6 J. Coyne, *Why Evolution is True*. New York: Viking Penguin, 2009. R. Dawkins, *The Greatest Show on Earth*. New York: Free Press, 2010.

7 R. Numbers, *The Creationists*. New York: Knopf, 1992. L. A. Witham, *Where Darwin Meets the Bible: Creationists and Evolutionists in America*. New York: Oxford University Press, 2002.

Chapter 1

1 J. Hutton, "Theory of the Earth; or an Investigation of the Laws Observable in the Composition, Dissolution, and Restoration of Land upon the Globe," *Earth and Environmental Science Transactions of The Royal Society of Edinburgh*, 1, 1788, pp. 209–304.

2 C. Darwin, *Journal of Researches into the Natural History and Geology of the Countries Visited During the Voyage of the H.M.S. Beagle Round the World, Under the Command of Capt. FitzRoy, R. N.*, 2nd ed., London: John Murray, 1845, p. 173. (This work is more popularly known as *The Voyage of the Beagle*, and Darwin added this sentence in the revised edition.)

3 Likewise the "catastrophism" of the French anatomist/paleontologist Georges Cuvier presupposed the origins of new species at different times in earth history.

4 H. Gee, *The Accidental Species: Misunderstandings of Human Evolution*, Chicago: University of Chicago Press, 2014, p. 105.

5 R. Song, "Play It Again, But This Time With Ontological Conviction," *Philosophy, Theology, and the Sciences*, 3, 2016, pp. 175–82, quotation from pp. 196–7.

6 J. F. Haught, *God After Darwin: A Theology of*

Evolution, 2nd ed., Boulder: Westview Press, 2007, p. xx, italics in original.

7 J. W. Van Huyssteen, *Alone in the World? Human Uniqueness in Science and Theology*, Grand Rapids: William B. Eerdmans, 2006, p. xviii.

8 R. Dawkins, *River Out of Eden*, New York: Basic Books, 1995, p. 131.

9 B. Latour, "Will Non-Humans be Saved? An Argument in Ecotheology," *Journal of the Royal Anthropological Institute*, 15, 2007, p. 470.

Chapter 2

1 T. Huxley, *Man's Place in Nature*, London: Williams and Norgate, 1863, p. 110.

2 E. Haeckel, *The History of Creation, Or the Development of the Earth and its Inhabitants by the Action of Natural Causes*, trans. E. R. Lankester, New York: D. Appleton, 1876, p. 365. Originally published in German as *Natürliche Schöpfungsgeschichte* (1868).

3 Unlike the human difficulty in parturition, menstruation is found across the monkeys and apes, and is sometimes known as "the curse" – although for no biblical reason.

4 See, for example, Jared Diamond, *The Third Chimpanzee*, New York: Harper, 1992.

5 The chromosomal process was discovered in the early twentieth century initially in female fruit flies, and is universally visible in reproductive cells under the light microscope as structures called "chiasmata."

Chapter 3

1 W. J. Bryan, "God and Evolution," *The New York Times*, February 26, 1922.

2 R. Virchow, "Prof. Virchow on Darwinism," *The Academy*, April 26, 1884, p. 299.

3 *New York Times Magazine*, March 16, 1916, pp. 4–5.

4 V. Kellogg, *Headquarters Nights*, Boston: The Atlantic Monthly Press, 1917, pp. 29–30.

5 "Full Text of Mr. Bryan's Argument against Evidence of Scientists," *The New York Times*, July 17, 1925, p. 2.

6 For example, Genesis 11: "When Arpachshad had lived thirty-five years, he became the father of Shelah; and Arpachshad lived after the birth of Shelah four hundred three years, and had other sons and daughters. When Shelah had lived thirty years, he became the father of Eber; and Shelah lived after the birth of Eber four hundred three years, and had other sons and daughters. When Eber had lived thirty-four years, he became the father of Peleg...," etc.

7 In 1970, i.e., two decades after *Worlds in Collision* and one decade after *The Genesis Flood*, Erich von Däniken's bestseller, *Chariots of the Gods?*, took on the field of archaeology, claiming that the Bible describes visits from extra-terrestrials whose true natures were misreported as works and visions of God. Von Däniken's ideas about "ancient astronauts" are now a mainstay of The History Channel on television. Approximately as many Americans believe in ancient astronauts as believe in creationism, and there is little information on just how

much overlap there may be between them, if any.

8 R. Dawkins, *The Blind Watchmaker*, New York: W. W. Norton, 1986, p. 54.

9 About the same time, Louisiana passed a "creation-science" law, which was eventually shot down by the Supreme Court in *Edwards v. Aguillard* (1987).

10 W. R. Overton, "Creationism in Schools: The Decision in McLean versus the Arkansas Board of Education," *Science*, 215, 1982, pp. 934–43, quotation from p. 939.

11 In the synoptic gospels, Jesus tells the parable of the wineskins: the vessel and its contents ought to match. Transferring or rebottling the contents is good neither for the wine nor the containers. Different translations render the containers as "wineskins" or "bottles."

12 "We Are Guarded by Spirits, Declares Dr. A. R. Wallace," *The New York Times*, October 8, 1911, p. 60.

13 R. Broom, "Evolution: Is There Intelligence Behind It?," *South African Journal of Science*, 30, 1933, pp. 1–19.

14 D. Hume, *Dialogues Concerning Natural Religion*, London (no publisher given), 1779, pp. 106–7.

15 The late law professor and designer of Intelligent Design, Phillip Johnson, answered decisively over lunch in 1999, after guest-lecturing in my class on human evolution at UC Berkeley. Species are like computer programs, Johnson had told the class, in particular, like Windows 98. When I pointed out that MacOS appeared to be the product of a different and superior designer, he readily acknowledged that Intelligent Design was not necessarily monotheistic.

Chapter 4

1 The word "literally" is sometimes ambiguous, as generations of Christian theologians have attempted to hew as closely to the biblical words as possible, while nevertheless recognizing that there are metaphors, contradictions, figures of speech, and flights of fancy in the words themselves. Consequently the challenge is always to uncover the meaning of the words.

2 For example, Isaiah 34:7, "And the unicorns shall come down with them, and the bullocks with the bulls; and their land shall be soaked with blood, and their dust made fat with fatness." Or Job 39:10, "Canst thou bind the unicorn with his band in the furrow? or will he harrow the valleys after thee?" Other translations give it as "wild ox." Some creationists indeed tether their beliefs to the existence of unicorns. See https://answersingenesis.org/extinct-animals/unicorns-in-the-bible.

3 Moreover, snakes are rather diverse, with thousands of species clustered into nearly two dozen families. If they all inherited the curse from this particular newly legless serpent, then since that time some of its descendants have certainly . . . diversified. Today some snakes are blind, some have no pelvic girdles, some are viviparous, some are parthenogenic, some are venomous, some are constrictors, some are specialized for life in water, others for life in sand, still others for life in trees. If that's not evolution, it sure is a lot of something else.

4 In his sermon at the wedding of Prince Harry and Meghan Markle, Bishop Michael Curry noted,

"French Jesuit Pierre Teilhard de Chardin was arguably one of the great minds, one of the great spirits of the twentieth century. A Jesuit, Roman Catholic priest, scientist, a scholar, a mystic. In some of his writings . . . he said as others have, that the discovery or invention or harnessing of fire was one of the great scientific and technological discoveries in all of human history. Fire to a great extent made human civilization possible. Fire made it possible to cook food and to provide sanitary ways of eating which reduced the spread of disease in its time. Fire made it possible to heat warm environments and thereby made human migration around the world a possibility, even into colder climates. Fire made it possible, there was no Bronze Age without fire, no Iron Age without fire, no industrial revolution without fire. The advances of science and technology are greatly dependent on the human ability and capacity to take fire and use it for human good."

5 That is, maize (UK) or mealie (South Africa).

Chapter 5

1 A Goldman Sachs analyst recently asked, "Is curing patients a sustainable business model?" CNBC, April 11, 2018, at https://www.cnbc.com/2018/04/11/goldman-asks-is-curing-patients-a-sustainable-busi ness-model.html.

2 "Ainsi, le cervelet du Huron contient en germe un esprit tout à fait semblable à celui de l'Anglais et du Français! Pourquoi donc, dans le cours des siècles, n'a-t-il découvert ni l'imprimerie ni la vapeur? Je serais en droit de lui demander, à ce Huron, s'il est égal à nos compatriotes, d'où il vient que les

guerriers de sa tribu n'ont pas fourni de César ni de Charlemagne, et par quelle inexplicable negligence ses chanteurs et ses sorciers ne sont jamais devenus ni des Homères ni des Hippocrates?" A. Gobineau, *Essai sur l'Inegalite des Races Humaines*, Paris: Firmin, 1853, pp. 60–1.

3 Darwin Correspondence Project, Letter no. 6684, Darwin to A. R. Wallace, March 27, 1869, at https://www.darwinproject.ac.uk/letter/DCP-LETT-6684.xml.

4 The next generation of anthropologists would de-synonymize culture and civilization, and redefine culture as something possessed completely, but differently, by all human societies.

5 The controversial 1962 book by Carleton Coon was titled *The Origin of Races*, invoking Darwin none too subtly, and was enthusiastically embraced by segregationists. The controversial 1994 book was *The Bell Curve*, co-authored by the reactionary hereditarian psychologist Herrnstein and conservative political theorist Charles Murray.

6 See C. Murray, "Deeper Into the Brain," *National Review*, January 24, 2000, at http://www.samtiden.com/tbc/las_artikel.php?id=19; and A. Sullivan, Denying Genetics Isn't Shutting Down Racism, It's Fueling It," *Intelligencer*, March 30, 2018, at http://nymag.com/daily/intelligencer/2018/03/denying-genetics-isnt-shutting-down-racism-its-fueling-it.html.

7 G. J. DeGroot, *The Bomb: A Life*, London: Jonathan Cape, 2005.

8 S. Pinker, *Enlightenment Now*, New York: Viking, 2018, p. 401.

9 See Meghan Henning, quoted in C. Moss, "The

Problematic Theology of 'Thoughts and Prayers,'"
Daily Beast, August 11, 2019, at https://www.
thedailybeast.com/the-problematic-theology-of-
thoughts-and-prayers; and M. Henning, "Prayer
Isn't a Gumball Machine," *The Christian Century*,
August 13, 2019, https://www.christiancentury.org/
blog-post/guest-post/prayer-isn-t-gumball-machine.

10 The basis lies in eschatology, not morality – in their
belief that their God wants them to tolerate or support immorality in the expectation of impending
heavenly rewards.

11 There are, however, important trickster figures in
the Bible, for example the patriarch Jacob, who similarly confounds his brother and rival; and, most
notably, the serpent.

12 On their 1971 album, *Every Good Boy Deserves
Favour*, The Moody Blues narrated history as a jour-
ney from the beginning of time to the end of time in
rhyme: "Desolation, Creation, Evolution, Pollution,
Saturation, Population, Annihilation, Revolution,
Confusion, Illusion, Conclusion, Starvation,
Degradation, Humiliation, Contemplation,
Inspiration, Elation, Salvation, Communication,
Compassion, Solution."

Chapter 6

1 The movie *2001: A Space Odyssey* famously cuts
directly from apes to space travel, leaving the actual
evolution of the human species completely unprob-
lematized.

2 R. A. Dart, "The Predatory Transition from Ape to
Man," *International Anthropological and Linguistic
Review*, 1, 1953, p. 209.

3 The "fruit of the Tree of the Knowledge of Good and Evil" became an "apple" as a Latin pun, where *malus* means both "evil" and "apple."

4 As are their symbolic relatives, human placentas.

5 But each ancestor is being counted multiple times, because we are all inbred – which is how we reconcile the astronomical number of one's lineal ancestors to the much smaller total number of people alive in any generation. This mathematical paradox is known as pedigree collapse.

6 D. B. Hart, *The Experience of God: Being, Consciousness, Bliss*, New Haven: Yale University Press, 2013, p. 28.

7 J. Barzun, *Science: The Glorious Entertainment*, New York: Harper and Row, 1964, p. 116.

8 "Cuinam quaeso sensum habenti consequenter videbitur dictum, quod dies prima et secunda et tertia, in quibus et vespera nominantur et mane, fuerint sine sole et sine luna et sine stellis; prima autem dies sine coelo? Quis vero ita idiotes invenietur, ut putet velut hominem quendam agricolam Deum plantasse arbores in paradiso; in Eden contra orientem, et arborem vitae plantasse in eo, id est, lignum visibile et palpabile, ita ut corporalibus dentibus manducans quis ex ea arbore, vitam percipiat, et rursus ex alia manducans arbore, boni et mali scientiam capiat? Sed et illud quod Deus post meridiem deambulare dicitur in paradiso, et Adam latere sub arbore, equidem nullum arbitror dubitare quod figurali tropo haec a Scriptura proferantur, quo per haec quaedam mystica indicentur" (Origen, *De Principiis* IV, 1, 16).

Bibliography

Alter, R., *Genesis*. New York: W. W. Norton, 1996.

Alumkal, A., *Paranoid Science: The Christian Right's War on Reality*. New York: NYU Press, 2017.

Anderson, W., *The Collectors of Lost Souls: Turning Kuru Scientists into Whitemen*. Baltimore: Johns Hopkins University Press, 2009.

Anonymous, "Full text of Mr. Bryan's argument against evidence of scientists: He contends that the law is clear, that it is proved that Scopes violated it and that the experts should have gone to the legislature," *The New York Times*, July 17, 1925.

Antón, S. C., Potts, R., and Aiello, L. C., "Evolution of Early Homo: An Integrated Biological Perspective," *Science*, 345:45, 2014.

Applegate, K. and Stump, J. B., eds., *How I Changed My Mind About Evolution: Evangelicals Reflect on Faith and Science*. Downers Grove: IVP Academic, 2017.

Bibliography

Armstrong, K., *In the Beginning: A New Interpretation of Genesis*. New York: Random House, 1997.

Arnold, B., "The Past as Propaganda: Totalitarian Archaeology in Nazi Germany," *Antiquity*, 64, 1990, pp. 464–78.

Augustine of Hippo, *On the Literal Interpretation of Genesis: An Unfinished Book*. In *Saint Augustine on Genesis*, trans. Roland J. Teske. Washington, DC: The Catholic University of America Press, 1991.

Avalos, H., "A Penis Bone in Genesis 2:21? Retrodiagnosis as a Methodological Problem in Scriptural Studies," at https://bibleinterp.arizona.edu/articles/penis-bone-genesis-221-retrodiagnosis-methodological-problem-scriptural-studies.

Bateman, A. J., "Intra-sexual Selection in Drosophila," *Heredity*, 2, 1948, pp. 349–68.

Bauer, H., *Beyond Velikovsky: The History of a Public Controversy*. Urbana: University of Illinois Press, 1984.

Beal, T., *The Rise and Fall of the Bible*. New York: Mariner, 2011.

Behe, M. J., *Darwin's Black Box: The Biochemical Challenge to Evolution*. New York: The Free Press, 1996.

Berkman, M. and Plutzer, E., *Evolution, Creationism, and the Battle to Control America's Classrooms*. New York: Cambridge University Press, 2010.

Bolnick, D. A., Fullwiley, D., Duster, T., Cooper, R. S., Fujimura, J., Kahn, J., Kaufman, J., Marks, J., Morning, A., Nelson, A., Ossorio, P., Reardon, J., Reverby, S., and Tallbear, K., "The Science and Business of Genetic Ancestry Testing," *Science*, 318, 2007, pp. 399–400.

Bibliography

Bowler, K., *Blessed: A History of the American Prosperity Gospel*. New York: Oxford University Press, 2013.

Bowler, P. J., *Monkey Trials and Gorilla Sermons: Evolution and Christianity from Darwin to Intelligent Design*. Cambridge, MA: Harvard University Press, 2007.

Brown, G. R., Laland, K. N., and Mulder, M. B., "Bateman's Principles and Human Sex Roles," *Trends in Ecology & Evolution*, 24, 2009, pp. 297–304.

Bryan, W. J., "God and Evolution," *The New York Times*, February 26, 1922.

Buckle, G. E., *The Life of Benjamin Disraeli, Earl of Beaconsfield, Volume IV: 1855–1868*. London: John Murray, 1929.

Buffon, Comte de, *Histoire Naturelle, Générale et Particuliére*, Vol. 4. Paris: Imprimerie Royale, 1753.

Bury, J. B., *The Idea of Progress: An Inquiry into its Origin and Growth*. New York: Macmillan, 1932.

Bynum, W. F., "Charles Lyell's *Antiquity of Man* and Its Critics," *Journal of the History of Biology*, 17, 1984, pp. 153–87.

Carsten, J., "Substance and Relationality: Blood in Contexts," *Annual Review of Anthropology*, 40, 2011, pp. 19–35.

Chapais, B., *Primeval Kinship*. Cambridge, MA: Harvard University Press, 2008.

Clark, K. J., *Religion and the Sciences of Origins*. New York: Palgrave Macmillan, 2014.

Darrow, C., "The Eugenics Cult," *The American Mercury*, 8, 1926, pp. 129–37.

Darwin, C., *Journal of Researches into the Natural History and Geology of the Countries Visited during*

the Voyage of H. M. S. Beagle Round the World,
under the Command of Capt. Fitz Roy, R. N., 2nd
edition. London: John Murray, 1845.

Dawkins, R., *The God Delusion*. London: Black Swan,
2006.

De Cruz, H. and Maeseneer, Y., "The Imago Dei:
Evolutionary and Theological Perspectives," *Zygon*,
49, 2014, pp. 95–100.

Deacon, T. W., *The Symbolic Species: The Co-evolution
of Language and the Brain*. New York: W. W. Norton
& Company, 1997.

Deane-Drummond, C., *Eco-Theology*. London: Darton,
Longman, and Todd, 2008.

Deane-Drummond, C. and Fuentes, A., eds., *Theology
and Evolutionary Anthropology: Dialogues in
Wisdom, Humility and Grace*. London: Routledge,
2020.

Dobzhansky, T., "Teilhard de Chardin and the
Orientation of Evolution," *Zygon*, 3, 1968,
pp. 242–58.

Ehrman, B., *Forged: Writing in the Name of God – Why
the Bible's Authors Are Not Who We Think They
Are*. New York: HarperCollins, 2011.

Eve, R. A. and Harrold, F. B., *The Creationist Movement
in Modern America*. Boston: Twayne, 1991.

Flannery, K. and Marcus, J., *The Creation of Inequality:
How Our Prehistoric Ancestors Set the Stage for
Monarchy, Slavery, and Empire*. Cambridge, MA:
Harvard University Press, 2014.

Forrest, B. and Gross, P. R., *Creationism's Trojan
Horse: The Wedge of Intelligent Design*. New York:
Oxford University Press, 2004.

Franklin, S. and McKinnon, S., eds., *Relative Values:*

Reconfiguring Kinship Studies. Durham, NC: Duke University Press, 2001.

Fuentes, A., *The Creative Spark: How Imagination Made Humans Exceptional*. New York: E. P. Dutton, 2017.

Fuentes, A., *Why We Believe: Evolution and the Human Way of Being*. New Haven: Yale University Press, 2019.

Gordin, M. D., *The Pseudoscience Wars: Immanuel Velikovsky and the Birth of the Modern Fringe*. Chicago: University of Chicago Press, 2012.

Gould, S. J., "Adam's Navel," *Natural History*, 93(6), 1984, pp. 6–14.

Gould, S. J., "Nonoverlapping Magisteria," *Natural History*, 106, 1997, pp. 16–22.

Graves, R. and Patai, R., *Hebrew Myths: The Book of Genesis*. New York: McGraw-Hill, 1963.

Greenblatt, S., *The Rise and Fall of Adam and Eve*. New York: W. W. Norton, 2017.

Greene, H. W., *Snakes: The Evolution of Mystery in Nature*. Berkeley: University of California Press, 1997.

Greene, H. W., "Evolutionary Scenarios and Primate Natural History," *The American Naturalist*, 190, 2017, S69–S86.

Gruber, J. W., "Brixham Cave and the Antiquity of Man." In *Context and Meaning in Cultural Anthropology*, ed. M. E. Spiro. New York: Free Press, 1965, pp. 373–402.

Haeckel, E., *The History of Creation: Or the Development of the Earth and its Inhabitants by the Action of Natural Causes*, trans. E. R. Lankester. New York: D. Appleton, 1868/1876.

Haeckel, E., *Natürliche Schöpfungsgeschichte*. Berlin: Reimer, 1868.

Haeckel, E. "Ernst Haeckel gives Germany's peace terms. Celebrated German scientist also discusses the probable effect of the present war upon social progress throughout the world," *The New York Times*, March 19, 1916.

Haught, J. F., *God After Darwin: A Theology of Evolution*, 2nd edition. Boulder: Westview Press, 2007.

Henke, W. and Tattersall, I., eds., *Handbook of Paleoanthropology* (online). Berlin: Springer, 2015.

Hertzel, D., *Ancestors: Who We Are and Where We Come From*. Lanham: Rowman & Littlefield, 2017.

Hesiod, *Theogony*, trans. M. L. West. New York: Oxford University Press, 1988.

Hofstadter, R., *Anti-Intellectualism in American Life*. New York: Knopf, 1963.

Hrdy, S. B., *Mothers and Others: The Evolutionary Origins of Mutual Understanding*. Cambridge, MA: Harvard University Press, 2009.

Huml, A. M., Sullivan, C., Figueroa, M., Scott, K., and Sehgal, A. R., "Consistency of Direct-to-Consumer Genetic Testing Results Among Identical Twins," *American Journal of Medicine*, 133, 2020, pp. 143–6.

Huxley, J., "At Random: A Television Preview." In *Evolution After Darwin, Vol. 3: Issues in Evolution*, ed. S. Tax and C. Callender. Chicago: University of Chicago Press, 1960, pp. 41–65.

Huxley, T., *Man's Place in Nature*. London: Williams and Norgate, 1863.

Insoll, T., "Ancestor Cults." In *The Oxford Handbook of the Archaeology of Ritual and Religion*, ed.

T. Insoll. New York: Oxford University Press, 2011, pp. 1043–58.

Jablonka, E. and Lamb, M. J., *Evolution in Four Dimensions: Genetic, Epigenetic, Behavioral, and Symbolic Variation in the History of Life*. Cambridge, MA: The MIT Press, 2005.

Jia Lanpo, *Early Man in China*. Beijing: Foreign Languages Press, 1980.

Jonas, H., *The Phenomenon of Life*. New York: Harper & Row, 1966.

Keel, T., *Divine Variations: How Christian Thought Became Racial Science*. Stanford: Stanford University Press, 2018.

Kellogg, V., *Headquarters Nights*. Boston: The Atlantic Monthly Press, 1917.

Kim, N. C. and Kissel, M., *Emergent Warfare in our Evolutionary Past*. New York: Routledge, 2018.

Kissel, M. and Fuentes, A., "From Hominid to Human: The Role of Human Wisdom and Distinctiveness in the Evolution of Modern Humans," *Philosophy, Theology and the Sciences*, 3, 2016, pp. 217–44.

Kistler, L., Maezumi, S. Y., Gregorio de Souza, J., Przelomska, N., Costa, F., Smith, O., Loiselle, H., Ramos-Madrigal, J., Wales, N., Ribeiro, E., Morrison, R., Grimaldo, C., Prous, A., Arriaza, B., Gilbert, M., Freitas, F., and Allaby, R., "Multiproxy Evidence Highlights a Complex Evolutionary Legacy of Maize in South America," *Science*, 362, 2018, pp. 1309–13.

Kottler, M. J., "Alfred Russel Wallace, the Origin of Man, and Spiritualism," *Isis*, 65, 1974, pp. 144–92.

Kuhn, T., *The Structure of Scientific Revolutions*. Chicago: University of Chicago Press, 1962.

Bibliography

Kuklick, H., ed., *A New History of Anthropology*. New York: Blackwell, 2008.

Laats, Adam, *Fundamentalism and Education in the Scopes Era: God, Darwin, and the Roots of America's Culture Wars*. New York: Palgrave Macmillan, 2010.

Larson, E. J., *Summer for the Gods: The Scopes Trial and America's Continuing Debate over Science and Religion*. New York: Basic Books, 1997.

Latour, B., "Will Non-Humans Be Saved? An Argument in Ecotheology," *Journal of the Royal Anthropological Institute*, 15, 2007, pp. 459–75.

Lilley, C. and Pederson, D., eds., *Human Origins and the Image of God*. Grand Rapids: William B. Eerdmans, 2017.

Liu, X., Lister, D. L., Zhao, Z., Petrie, C. A., Zeng, X., Jones, P. J., Staff, R. A., Pokharia, A. K., Bates, J., Singh, R. N., Weber, S. A., Matuzeviciute, G. M., Dong, G., Li, H., Lü, H., Jiang, H., Wang, J., Ma, J., Tian, D., Jin, G., Zhou, L., Wu, X., and Jones, M. K., "Journey to the East: Diverse Routes and Variable Flowering Times for Wheat and Barley En Route to Prehistoric China," *PLoS ONE*, November 2, 2017.

Livingstone, D., *Adam's Ancestors: Race, Religion, and the Politics of Human Origins*. Baltimore: Johns Hopkins University Press, 2008.

Livingstone, D. N., *Dealing with Darwin: Place, Politics, and Rhetoric in Religious Engagements with Evolution*. Baltimore: Johns Hopkins University Press, 2014.

McGrath, A. E., *Darwinism and the Divine: Evolutionary Thought and Natural Theology*. New York: John Wiley & Sons, 2013.

Marks, J., *Tales of the ex-Apes: How We Think about*

Human Evolution. Berkeley: University of California Press, 2015.

Marks, J., "Genetic Testing: When Is Information Too Much?," *Anthropology Today*, 34, 2018, pp. 1–2.

Meloni, M., *Political Biology: Science and Social Values in Human Heredity from Eugenics to Epigenetics*. New York: Palgrave Macmillan, 2016.

Montagu, M. F. A., "Edward Tyson, M.D., F.R.S., 1650–1708 and the Rise of Human and Comparative Anatomy in England," *Memoirs of the American Philosophical Society*, 20, 1943, pp. 1–488.

Nelkin, D. and Lindee, M. S., *The DNA Mystique: The Gene as Cultural Icon*. New York: Freeman, 1995.

Nisbet, R., *History of the Idea of Progress*. New York: Basic Books, 1980.

Numbers, R. L., *The Creationists*. New York: Knopf, 1992.

Overton, W. R., "Creationism in Schools: The Decision in McLean versus the Arkansas Board of Education," *Science*, 215, 1982, pp. 934–43.

Panofsky, A. and Donovan, J., "Genetic Ancestry Testing Among White Nationalists: From Identity Repair to Citizen Science," *Social Studies of Science*, 49, 2019, pp. 653–81.

Parsons, Elsie Clews, "Links Between Religion and Morality in Early Culture," *American Anthropologist*, 17, 1915, pp. 41–57.

Pennock, R. T., "Creationism and Intelligent Design," *Annual Review of Genomics and Human Genetics*, 4, 2003, pp. 143–63.

Petto, A. J. and Godfrey, L., eds., *Scientists Confront Intelligent Design and Creationism*. New York: W. W. Norton, 2007.

Pickering, T. R., *Rough and Tumble: Aggression, Hunting, and Human Evolution*. Berkeley: University of California Press, 2013.

Pobiner, B., "Accepting, Understanding, Teaching, and Learning (Human) Evolution: Obstacles and Opportunities," *Yearbook of Physical Anthropology*, 159, 2016, S232–S274.

Porr, M. and Matthew, J., eds., *Interrogating Human Origins: Decolonisation and the Deep Human Past*. London: Routledge, 2020.

Preuss, T. M., "The Human Brain: Evolution and Distinctive Features. In *On Human Nature: Biology, Psychology, Ethics, Politics, and Religion*, ed. M. Tibayrenc and F. J. Ayala. London: Elsevier, 2016, pp. 125–50.

Re Manning, R., ed., *The Oxford Handbook of Natural Theology*. New York: Oxford University Press, 2013.

Reverby, Susan M., "Ethical Failures and History Lessons: The US Public Health Service Research Studies in Tuskegee and Guatemala," *Public Health Reviews*, 34, 2012, pp. 13–30.

Richmond, J., "Design and Dissent: Religion, Authority, and the Scientific Spirit of Robert Broom," *Isis*, 100, 2009, pp. 485–504.

Ridley, M., *The Rational Optimist: How Prosperity Evolves*. New York: HarperCollins, 2010.

Robbins, R. H. and Cohen, M. N., eds., *Darwin and the Bible: The Cultural Confrontation*. Boston: Pearson Education, 2009.

Rudwick, M. J. S., *Worlds Before Adam: The Reconstruction of Geohistory in the Age of Reform*. Chicago: University of Chicago Press, 2008.

Bibliography

Ruse, M., *The Evolution-Creation Struggle*. Cambridge, MA: Harvard University Press, 2005.

Ruse, M., *The Problem of War: Darwinism, Christianity, and Their Battle to Understand Human Conflict*. New York: Oxford University Press, 2019.

Russell, Nerissa, *Social Zooarchaeology: Humans and Animals in Prehistory*. New York: Cambridge University Press, 2011.

Rutherford, A., *A Brief History of Everyone Who Ever Lived*. London: Weidenfeld and Nicolson, 2016.

Sacks, J., *The Great Partnership: Science, Religion, and the Search for Meaning*. New York: Schocken, 2011.

Saini, A., *Inferior: How Science Got Women Wrong – and the New Research That's Rewriting the Story*. Boston: Beacon, 2017.

Saini, A., *Superior: The Return of Race Science*. Boston: Beacon, 2019.

Schloss, J. P., "Evolutionary Theory and Religious Belief." In *The Oxford Handbook of Religion and Science*, ed. P. Clayton. New York: Oxford University Press, 2006, pp. 187–206.

Schmalzer, S., *The People's Peking Man: Popular Science and Human Identity in Twentieth-Century China*. Chicago: University of Chicago Press, 2008.

Scott, E. C., *Evolution vs. Creationism: An Introduction*. Berkeley: University of California Press, 2005.

Shea, J. J., *Stone Tools in Human Evolution: Behavioral Differences Among Technological Primates*. New York: Cambridge University Press, 2016.

Skloot, R., *The Immortal Life of Henrietta Lacks*. New York: Crown, 2010.

Slack, G., *The Battle over the Meaning of Everything:*

Evolution, Intelligent Design, and a School Board in Dover, PA. San Francisco: Jossey-Bass, 2007.

Sloan, P. R., "The Buffon-Linnaeus Controversy," *Isis*, 67, 1976, pp. 356–75.

Song, R., "Play It Again, But This Time With Ontological Conviction," *Philosophy, Theology, and the Sciences*, 3, 2016, pp. 175–82.

Stocking, George W., *After Tylor: British Social Anthropology, 1888–1951*. Madison: University of Wisconsin Press, 1998.

Strier, K. B., *Primate Behavioral Ecology*, 5th edition. New York: Routledge, 2017.

Tabor, J. *The Book of Genesis*. Charlotte: Genesis 2000, 2020.

Tatje, T. and Hsu, F. L. K., "Variations in Ancestor Worship Beliefs and Practices," *Southwestern Journal of Anthropology*, 25, 1969, pp. 153–72.

Teilhard de Chardin, P., *The Phenomenon of Man*. New York: Harper & Row, 1959.

Trevathan, W. R. and Rosenberg, K. R., *Costly and Cute: Helpless Infants and Human Evolution*. Albuquerque: University of New Mexico Press, 2016.

Turner, J., *Philology: The Forgotten Origins of the Modern Humanities*. Princeton: Princeton University Press, 2014.

Tylor, Edward B., *Primitive Culture: Researches into the Development of Mythology, Philosophy, Religion, Art, and Custom*. London: John Murray, 1871.

Ungar, P. S., *Evolution's Bite: A Story of Teeth, Diet, and Human Origins*. Princeton: Princeton University Press, 2017.

Van Arsdale, A. P., "Population Demography, Ancestry,

and the Biological Concept of Race," *Annual Review of Anthropology*, 48, 2019, pp. 227–41.

Van Huyssteen, J. W., *Alone in the World? Human Uniqueness in Science and Theology*. Grand Rapids: William B. Eerdmans, 2006.

Van Riper, A., *Men Among the Mammoths: Victorian Science and the Discovery of Human Prehistory*. Chicago: University of Chicago Press, 1993.

VanderKam, J., *The Book of Jubilees*. Sheffield: Sheffield Academic Press, 2001.

Virchow, R., "The Liberty of Science in the Modern State, III," *Nature*, 17, 1877, pp. 111–13.

Virchow, R., "Prof. Virchow on Darwinism," *The Academy*, April 26, 1884, p. 299.

Wallace, Alfred R., [Review of Sir Charles Lyell on geological climates and the origin of species], *Quarterly Review*, 126, 1869, pp. 359–94.

Walton, J. H., *The Lost World of Adam and Eve*. Downers Grove: InterVarsity Press, 2015.

Washington, Harriet A., *Deadly Monopolies*. New York: Random House, 2012.

Whitehead, A. N., *Process and Reality*. New York: Macmillan, 1929.

Wrangham, R., *Catching Fire*. Cambridge, MA: Harvard University Press, 2009.

Young, S. and Benyshek, D., "In Search of Human Placentophagy: A Cross-Cultural Survey of Human Placenta Consumption, Disposal Practices, and Cultural Beliefs," *Ecology of Food and Nutrition*, 49(6), 2010, pp. 467–84.

Zevit, Ziony, *What Really Happened in the Garden of Eden?* New Haven: Yale University Press, 2013.

Index

Index

apes (*cont.*)
  dental features 24, 27
  descent of man from xi,
    18–21, 22, 43, 44, 54, 57,
    104–5, 113
  differentiation between
    humans and 22–4, 27–8,
    32
  giving birth 28–9
  similarities to humans 22
Aquinas, Thomas 59
Aristotle 3
Arkansas
  suing of over creation science
    law 55
art, prehistoric 33–4
astronauts, ancient 53, 122
astronomy xii, 52
  use of by Velikovsky in
    interpretating Bible stories
    xii–xiii, 51–3
atomic bomb 90–1
Augustine, St 59, 68, 75
Australopithecus 2, 24, 25–6,
    30, 106, 107
  afarensis 2
  sediba 2

baboons 39
Bacon, Francis 19
Bakker, Jim 95
Barzun, Jacques 114–15
Bastian, Adolf 44–5
behavioral humanity 32–3
Berlin Anthropological Society
    44–5
Bible xiv, 3–4, 17, 49
  citing of non-existent source
    70–1
  and classification of animals
    48
  and cultural evolution 75–80
  Garden of Eden/Adam and
    Eve 4–5, 12, 13, 28, 75,
    98, 99

and geometry 72
inaccuracies and
  inconsistencies 65–74
sacredness of 12, 102
similar chapters in different
  places 71–2
Velikovsky's interpretation of
  stories through astronomy
  xii–xiii, 51–3
view of by creationists 94
biblical creationism 4–5, 9, 10
biblical literalism/literalist xi,
  xiii, 56, 58, 62, 116
difference between
  kleptomaniac and ix
and rationalism 65–80
and science ix
biblical literalist creationism
  xiii, 9, 54, 115
biblical rationalism 51, 71–4
bipedalism 23, 24, 26
brain 26, 28, 35, 84
  expansion of 31
  explaining group differences
  86
British Association for the
  Advancement of Science 11
Broom, Robert 57
Bryan, William Jennings 42–3,
  46–9, 50, 86
Buffon, Count de 20, 21
burning, development of 29–30

Cain and Abel 76, 77–8
cave paintings 33
childbirth 28–9, 108
chimpanzees 20, 29, 76
Christian fundamentalism xi
Christianity
  and Darwinism 17–18
  and evolution 14–18
Cicero 19
civilization
  nature and attainment of
    13–14

Index

Index

Index

Index

Index

science
  and biblical literalists ix
  and creationists xiii, 56,
    93–4, 102–3, 113
  fad 87
  and humanities 115
  and inequality 82–3
  and the moral realm 97–103
  rejection of *see* anti-science
    movement
  and religion xiv, 14, 89–93,
    114
  and technology 81
Scopes Trial 37, 46–50, 55
sexual dimorphism 39–40
sexual evolutionary processes
  39–40
sexual selection 86
siblings 32
Simpson, George Gaylord 16
slaves-slavers relationship 12
Smith, Adam 62
  The Wealth of Nations 62
social Darwinism 46
social relationships
  development of by humans
    32–3
Sollas, William J.
  Ancient Hunters x
Solomon, King 72
sound production (speech) 26–7
species
  extinction of 7–8
  as machinery/subspecies
    debate 79–80
  transmutation of 8
spirituality/spiritualism 20, 57,
  58, 63, 85, 109
Star Wars 106
sterilization of poor people 1,
  37, 43, 113
Stone Age 13
stone tools 29, 77
Sullivan, Andrew 89
symbolic thought 24

taboos 31
Tatian 68
technology
  as engine of civilization 82
  improvement of 81
  and science 81
teleological theory, of evolution
  15–16
Teilhard de Chardin, Pierre
  15–16
Ten Plagues of Egypt xii, 51–2
teosinte 78
Texas 60
theology xiv, 75, 94, 102, 114
  and evolution 14–18, 55
  natural 7, 9–11, 59, 62, 79
thermodynamics, second law
  of 54
tools 29, 77
Truman, Harry 90–1
Trump, Donald 96
Tuskegee syphilis experiments
  92
23andMe 109–10
Tylor, Edward 85–6, 89
  Primitive Culture 85–6
Tyson, Edward 19–20

uncivilized peoples 82–3
uniformitarianism 7
Ussher, James 49

vaccines
  opposition to by Americans
    94
Velikovsky, Immanuel
  Worlds in Collision xii–xiii,
    51–2
Venus 51–2
Vestiges of Creation 8
Victorian creationism 9
Virchow, Rudolf 44–5

Wallace, Alfred Russel 57, 84,
  85

Index